The Rightful Place of Science:
Politics

The Rightful Place of Science:
Politics

Edited by

G. Pascal Zachary

Contributors

Michael Crow
Robert Frodeman
David Guston
Carl Mitcham
Daniel Sarewitz

Consortium for Science, Policy and Outcomes
Tempe, AZ and Washington, DC

Model citation for this volume:

Zachary, G.P., ed. 2013. The Rightful Place of Science: Politics. Tempe, AZ: Consortium for Science, Policy and Outcomes.

The Rightful Place of Science series explores the complex interactions among science, technology, politics and the human condition.

Other volumes in this series:

Alic, John A. 2013. The Rightful Place of Science: Biofuels. Tempe, AZ: Consortium for Science, Policy and Outcomes.

ISBN: 0615886701

ISBN-13: 978-0615886701

LCCN: 2013919964

First Edition, November 2013

CONTENTS

EDITOR'S NOTE

G. Pascal Zachary

The rapidity of technological and scientific change—the novelty of the new, its seemingly infinite imperatives for reshaping how humans live and dream, the vast unexplored territory of human understanding that awaits further exploration—makes serious attempts to understand the nature of technological and scientific change seemingly difficult, if not flatly impossible. Techno-utopians insist, Don't worry, be happy. *Uber*-scientists and engineers insist: stay out our way, we know best. Their perspective can be persuasive; after all, many digital devices do so many amazing things that consumers can easily become slavish enthusiasts. Amidst such technological and scientific abundance, who drives change and to what end are questions that not only elude us but often are dismissed as irrelevant.

Some people insist our technologies deliver the best of all possible worlds. They raise doubts that greater understanding of technological change won't do anything other than make us more aware of its inevitability. What might we do differently, anyway, these skeptics say, even if we do work to understand the genuine sources of technological change?

In this volume, we gather a collection of thinkers who insist there is much to gain from trying to comprehend the *politics* of technological change and, its close cousin, the practice of science and scientific research. The authors are part of an intellectual and ethical movement to normalize science and technology: to view these activities neither as objects of worship nor mere scholarly analysis. The authors are keen, close observers of the science scene, but significantly they have a stake in the game. They wish to improve on the politics of science and to judge their reforms by a pragmatic measure: by the outcomes of science and technology.

I must be careful here not to pre-empt the sharp insights and pithy discussions of hard problems and elusive processes that await the diligent reader. But let me give a few hints at what's to come. First, these authors share a world-view that is deftly summarized in the opening essay by Michael Crow and Daniel Sarewitz:

"In a world where scientific research and technological innovation receive, on the whole, more than a trillion dollars of support from government and industry, the idea that the advance of knowledge and innovation can be understood separately from the political, institutional, cultural and economic contexts within which science and technology are advanced is nonsense. We'll go further: given the powers we can wield with our scientific and technological products, the idea that we cannot and should not do a better job exercising wise governance over the contexts of their creation and use is not just irresponsible, it is likely to be destructive."

Easy answers and simple formulas are not the diet these authors demand. Instead, they see paradox and complexity at every turn, and they are unafraid to squarely face the hunger many of us feel for something *more* from our enormous civilization expenditures on science and engineering. In these essays, which are meant in many ways as a primer for the uninitiated and a stimulus for the already-initiated, there is always pith, and clarity, nuance and vision, as when Crow writes in chapter three: "Too often we choose technologies that save us from today's predicament but add to the problems of tomorrow." And a sense of authority and considered judgment, as when David Guston observes, in chapter four, "Science is deeply political and always will be." Or when Sarewitz, eschewing easy solutions, confesses in the volume's ultimate essay, that "the rightful place of science, in short, is hard to find," but that the search is well worth undertaking and will reward those who do.

To these authors, how we talk about technological change matters, because policies ultimately express deeper vernacular yearnings—for democracy, equity and of course utility. Through these essays, hard questions get asked, new perspectives are presented and the impulse towards contrarian understandings—towards going against the herd—abound. So when Robert Frodeman and Carl Mitcham conclude in the elegant reconsideration of the "social contract" and science, "Neither scientist nor citizens live by contract alone," we know that we are traveling the necessary intersection between revision and reform. And when Sarewitz declares, in a trenchant and revelatory essay on budgetary politics and science, "discussions are dominated by concerns about 'how much' and avoid like the plague serious questions about 'what for,' he knows that spending is no proxy for values, and that scientific and technological history is chock full of examples of *lesser funded* efforts carrying the day.

All five of the authors of *The Rightful Place of Science: Politics* are closely associated with the Consortium for Science Policy & Outcomes at Arizona State University. Two of the authors are co-founders of CSPO; another is currently a co-director; all contributors are prodigious scholars in their own right. I am greatly privileged to work myself at CSPO and ASU, as a professor of practice. I also am pleased to serve as founding editor of

CSPO's publishing series, The Rightful Place of Science. We have taken the name from a phrase uttered by Barack Obama in his first presidential inaugural in January 2009. To restore science to its rightful place can have multiple meanings, and in this volume, some of those meanings will be explored by thinkers far more perspicacious than I. Let me only suggest here that establishing the proper relationship between humans and their techno-scientific practices is a task that stands beyond any particular presidential administration or even national polity. Thinking, feeling people throughout the planet are condemned to confront the unfinished task of defining and redefining "the rightful place of science" in their own lives, in their own societies — and for their own futures. In the two centuries since Darwin and Marx radically altered the relationship between man and nature, on the one hand, and man and his tools, on the other hand, the human species has wrestled with the growing sense that somehow both biology and technology are out of control. To comprehend the essential tension between the natural and the artificial remains an animating concern for any movement that claims to both understand techno-scientific change — and influence its hydra-headed course.

I come to the task with two significant endowments. First I am the author of a comprehensive biography of Vannevar Bush, whom we might consider the first politician of science and technology. Bush authored *Science, the Endless Frontier,* a report to the President in the waning months of World War II that set the terms of engagement between science and government, research and society for a half-century. Second, as a journalist and essayist, chiefly for *The Wall Street Journal, The New York Times* and *Technology Review*, I spent two decades closely following Silicon Valley, the world's premier innovation enclave and unquestionably the outstanding provider of evidence for U.S. world leadership in science and technology

In this volume, and in future volumes, we seek to present complex and compelling analyses of many aspects of techno-scientific change in vernacular language. Our perspective is pragmatic and future-oriented. We shall often publish brief introductions to complicated subjects, knowing full well the risks of leaving advanced readers unsatisfied. Leveraging new inexpensive publishing tools, we insist on joining the revolution in serious scholarly publishing that is already well underway. Global in scope, the rise of e-books and print-on-demand are technologies that will bring serious scholarship on science, technology and society to new audiences. While eschewing the aim of comprehensiveness, The Rightful Place of Science series will strive instead to engage the curious, stimulate the well-informed and challenge those with steadfast positions and perspectives.

1

POWER AND PERSISTENCE
IN THE POLITICS OF SCIENCE

Michael Crow and Daniel Sarewitz

Science and technology are extraordinarily powerful forces of social change in our world. But the implications of this commonplace observation are poorly understood — and not just by citizens, but, equally, by the scientists, technologists, and government and industry leaders who imagine they are guiding our efforts to apply knowledge and innovation for human benefit.

The human prospect, in short, is linked to an endeavor so deeply ambiguous that it is not only essential for our survival, but is also the potential source of our self-destruction. And yet, if we consider what we know about how to govern our undeniably enormous power to understand and manipulate matter and nature, we are at best in a stage of primitive understanding, at worst in a state of congenital denial. We are moving toward warfare fought by remote controlled and increasingly autonomous robotic weapons with little consideration of what the consequences will be for international rule of law, not to mention for our own long-term security. We accelerate the growth of new, high-tech industries that don't seem to require much in the way of a skilled workforce even as they create untold wealth for a tiny percent of our populace. We pour billions of dollars into efforts to understand cancer and climate change without much advancing our ability to prevent the suffering that they cause. We tout the coming revolution in personalized genomic medicine without acknowledging that many of our most serious health problems are rooted in society, culture, the economy, the environment, and even the medical system itself — rather than in our genes. We proudly describe the economic consequences of technological change as "creative destruction," conveniently ignoring that what gets destroyed is not just old industries but also people's jobs and communities, and even the social fabric itself. The fact is that human thriving and human suffering are both the consequences of scientific and technological change, and if the ledger books over centuries have shown a net gain for thriving, this is not a natural law, and neither will it much matter to those on the losing end.

The stakes of this gambit between thriving and suffering are higher today than they have ever been, and growing higher by the day. The globalization of economies and the technological empowerment of individuals deliver an ever increasing complex of opportunity (as, for example, hundreds of millions of people are lifted from poverty through the wealth-creating power of technological change) and danger (as, for example, even one of the poorest nations on Earth, North Korea, can manage to build nuclear weapons, while soon anyone with a biology degree may well be able to create lethal pathogens in their garage). The elimination of any meaningful boundary between the artificial and the natural (the Earth's atmosphere and biota must both now be understood as in part manufactured) gives rise to new levels of complexities that continually evade our technological and political imaginations—for example, consider that in New York City, some of the newest, most energy efficient and "sustainable" buildings turned out to be the most vulnerable to Hurricane Sandy because their highly sophisticated technological infrastructures were located in the basements. (Never mind that the inevitability of an event like Sandy was entirely evident decades before it actually occurred.)

Intellectually serious and concerted efforts to understand the complex interdependencies among science, technology, society, and nature date back only to the middle of the 20th century, and to such thinkers as Lewis Mumford and J.D. Bernal, Rachel Carson and Bertrand Russell. Efforts to systematically organize inquiry and deliberation about such problems are younger still, and represent at best a marginal commitment in our academic, private sector, and government institutions. Awareness of the importance of these issues has grown in some corners of the policy world, but can hardly be said to have achieved even a tiny fraction of the attention that their importance would seem to merit. Most science journalism is little more than a mouthpiece for the press releases of university and industry PR operations; political attention is little more than a focus on budget levels, or easily-exploited, emotionally-charged issues like embryonic stem cells and climate change; while the science community typically offers little more than demands for more money and more unquestioning faith in its goodness. These are sweeping generalizations, of course, but much closer to a true characterization of the situation than one would wish. As just one small example, when the first author of this chapter deigned to publish in the international science journal *Nature* a short commentary—a thought experiment, in fact—on how we might re-conceptualize our biomedical research enterprise to generate more positive health outcomes, the ideas was treated by many, including leaders of the scientific community, as if it was attacking science itself—when in reality it was simply asking that we be willing to question the status quo on behalf of society.

Long ago, Thomas Kuhn taught us that science only changes with the generations, and only when the degree of incoherence of our ideas becomes so inescapable that new paradigms are required for understanding what is going on. So resistance to discussions about change— not to mention to change itself—ought to be expected. Yet we do remain surprised at how difficult it has proven to seriously ask difficult questions about how we are to manage our growing scientific and technological prowess in the face of the consequent growth of complexity, ambiguity, and potential for great gain and great harm alike.

We can locate much of this difficulty in a set of powerful cultural beliefs that have little empirical support, but much political and psychological value as sources of identity not just among scientists and technologists, but in modern Western societies more broadly. Chief among these is the comforting but utterly unhelpful idea that science and technology are somehow independent of culture and politics. In a world where scientific research and technological innovation receive, on the whole, more than a trillion dollars of support from government and industry, the idea that the advance of knowledge and innovation can be understood separately from the political, institutional, cultural, and economic contexts within which science and technology are advanced is nonsense. We will go further: given the powers that we can wield with our

4

scientific and technological products, the idea that we cannot and should not do a better job exercising wise governance over the contexts of their creation and use is not just irresponsible, it is likely to be self-destructive.

Not only do we cling to the idea that science and technology are "autonomous," but we buttress this notion with a set of subsidiary ideas that are equally simplistic and distracting: that science is either "basic or applied;" that science is simply "translated" into action; that technology derives from science, or even that the two can be meaningfully separated; that scientific and technological advance always offer improvements over past practices; that technologies are either risky or beneficial, or perhaps worse, that the ratio of risk to benefit can be quantified in a meaningful way; that if only the public were "scientifically literate" our society would suddenly achieve some sort of Elysian rationality. These sorts of dichotomies and simplistic notions are shackles on our minds, our imaginations, and our public deliberations; they restrain us from authentically encountering and engaging the daunting complexity of our scientific and technological endeavor.

As one very modest effort to try to create some intellectual space for developing alternatives to these powerful but primitive notions, the Center for Science, Policy and Outcomes was founded at Columbia University in 1999. The center was subsequently rechristened the Consortium for Science, Policy and Outcomes in 2004, shortly after the first author assumed the Presidency at Arizona State University, the new name meaning to signal a broader ambition to build a social and institutional capacity that might begin to come to terms with the challenges of wisely governing our knowledge and innovation enterprises. We don't mean for a minute to suggest that CSPO is the only effort in this domain, or even the most effective or most important, only that we have been committed to bringing out into the open momentous and difficult problems — while hopefully also pointing the way to creative and inclusive solutions — that seem astonishingly absent from the public sphere.

It would of course be ridiculous to say that the contents of this, or any book, represented a major step away from our intellectual and moral infantilism about such matters toward a responsible maturity, but at least we are trying to throw down the gauntlet. The most important thing to be done at this stage is to begin to map out the landscape that we must navigate — a terrain that grows more complex and incomprehensible as our scientific understanding and technological prowess continues to expand.

Thus, the eight essays collected in this book should be seen as small pieces of a large map that we are assembling along with many others. The goal of the map is to allow a better orientation, a clearer picture of the options open to society, and the potential obstacles and opportunities created by those options, as our society continues to pursue its commitment to scientific and technological change. The subjects of the chapters vary, as they must, from high-level reflections about the limits of human understanding, to on-the-ground suggestions about how resources for science and technology might best be allocated at the federal level, or technologies advanced at the state level. But we mean them to add up to a sweeping introduction to the problem and the challenge created by humanity's unique genius to discover and create. In future volumes we will revisit the issues raised here, delving into them in greater depth and with a greater diversity of perspectives and aims.

It is a tribute to his vision, and an indictment of our own, that words spoken by President Eisenhower more than half a century ago retain their freshness and urgency today in capturing the challenges at hand: "Man's power to achieve good or to inflict evil surpasses the brightest hopes and the sharpest fears of all ages. We can turn rivers in their courses, level mountains to the plains. Oceans and land and sky are avenues for our colossal commerce. Disease diminishes and life lengthens. Yet the promise of this life is imperiled by the very genius that has made it possible. Nations amass wealth. Labor sweats to create, and turns out devices to level not only mountains but also cities. Science seems ready to confer upon us, as its final gift, the power to erase human life from this planet." It is time to take very seriously not just the promise, but also the warning, conveyed in these words.

2

MAKING SCIENCE POLICY MATTER FOR A USE-INSPIRED SOCIETY

Daniel Sarewitz

It is axiomatic and also true that federal science policy is largely played out as federal science budget policy. Science advocacy organizations such as the American Association for the Advancement of Science (AAAS), the National Academies, and various disciplinary professional societies carefully monitor the budget process and publish periodic assessments, while issue-focused interest groups such as disease lobbies and environmental organizations focus on agencies and programs of specific relevance to their constituencies. Overall, it is fair to say that marginal budgetary changes are treated by the science and technology community as surrogates for the well-being of the science enterprise, while the interested public considers such changes to be surrogates for progress toward particular societal goals (for example, budget increases for cancer research mean more rapid progress toward cures). In this dominant science policy worldview, yearly budget increases mean that science is doing well, and doing good.

When budgets are flat or declining — or even when rates of budget increases are slowing — then science must suffer and so, by extension, must the prospects for humanity. In recent years, this worldview was perhaps most starkly on display in discussions about the National Institutes of Health (NIH), whose budget doubled between 1998 and 2003 as a result of a highly effective lobbying effort, a sympathetic Congress, and a brief period of overall budgetary surplus. During this period, the NIH budget went from $13.6 billion to $27.0 billion, and the NIH share of all civilian federally funded research rose from its already dominant 37% to 48%. Nevertheless, when the fiscal year (FY) 2004 budget debates began, NIH and its advocates in the research community portrayed the situation as one of crisis arising from a sudden decline in the rate of budget increase. Said a representative of the Association of American Medical Colleges: "Two or three years of 2 or 3% increases, and you've pretty much lost what you've gained . . . And you've certainly lost the morale of investigators who can't help but be demoralized by trying to compete for funding under those circumstances." In another notable example, the president-elect of the AAAS in 1990 solicited letters from 250 scientists and discovered that many were unhappy because they felt that they did not have enough funding. From this information he inferred an "impoverishment" of basic research even though, as he acknowledged, science funding had been growing steadily.

In this article I will argue that the annual obsession with marginal changes in the research and development (R&D) budget tells us something important about the internal politics of science, but little, if anything, that's useful about the health of the science enterprise as a whole. In particular, marginal budget changes give almost no information about the capacity of the science enterprise to contribute to the wide array of social goals that justifies society's investment in science. I will return to this point later; first I will focus on the more parochial reality that the annual federal budget numbers for science cannot be understood unless they are placed in a broader political and historical context. A given year's marginal budget increase says as much about the health of the science enterprise as the nutritional value of a single meal says about the health of one's body.

One of the most astonishing aspects of science policy over the past 30 or so years is the consistency of R&D funding levels as a proportion of the discretionary budget. (Discretionary spending is the part of the budget that is subject to annual congressional decisions about spending levels.) Since the mid-1970s, nondefense R&D budgets have constituted between 10 and 12% of total nondefense discretionary spending. Total R&D (defense and non-defense) shows a similar stability at 13 to 14% of the total discretionary budget. This consistency tells us that marginal changes in the R&D budget are tightly coupled to trends in discretionary spending as a whole.

Given the Balkanized manner in which science budgets are determined, such stability at first blush may seem incomprehensible. After all, no capacity exists in the U.S. government to undertake centralized, strategic science policy planning across the gamut of federal R&D agencies and activities. The seat of U.S. science policy in the executive branch is the Office of Science and Technology Policy, whose director is the president's science advisor. The influence of this position has waxed and waned (mostly waned) with time, but it has never been sufficient to exercise significant control over budgetary planning. That control sits with the Office of Management and Budget, which solicits budgetary needs from the many executive agencies that conduct R&D, negotiates with and among the agencies to reach a final number that is consistent with the president's budgetary goals, and then combines the individual agency budgets for reporting purposes into categories that create the illusion of a coherent R&D budget. But this budget is an artificial construct that conceals the internal history, politics, and culture of each individual agency.

The situation in Congress is even more Byzantine, with 20 or more authorization and appropriations committees (and innumerably more subcommittees) in the Senate and House each exercising jurisdiction over various pieces of the publicly funded R&D enterprise. Moreover, the jurisdiction of the authorizing committees does not match that of the appropriations committees; nor do the allocations of jurisdiction among Senate committees match those of the House. Finally, the appropriations process puts science and technology (S&T) agencies such as the National Science Foundation (NSF) and the National Aeronautics and Space Administration (NASA) in direct competition with other agencies such as the Department of Justice and the Office of the U.S. Trade Representative for particular slices of the budgetary pie.

The decentralization of influence over S&T budgeting in the federal government precludes any strategic approach to priority setting and funding allocations. Although an "R&D" budget can be—and is—constructed and analyzed each year, this budget is an after-the-fact summation of numerous independent actions taken by congressional committees and executive-branch bodies, each of which is in turn influenced by its own set of constituents and shifting priorities. From this perspective, if science policy is mostly science budget policy, then one can reasonably assert that there is no such thing as a national science policy in the United States.

If central science policy planning in the United States is impossible, how is one to make sense of the remarkable stability of R&D spending as a proportion of the total discretionary budget? Several related factors come into play.

First, the political dynamics of budget-making result in a highly buffered system where every major program is protected by an array of advocates and entrenched interests fighting for more resources and thus, on the whole, offsetting the efforts of other advocates and interests trying to advance other programs.

Second, annual marginal changes in any program or agency budget are generally small: This year's budget is almost always the strongest predictor of next year's budget. Large changes mean that a particular priority has gained precedence over other, competing ones, and such situations are not only uncommon but usually related to a galvanizing political crisis, such as 9/11, or the launching of Sputnik.

Third, in light of the previous considerations, annual changes in expenditure levels, whether for R&D programs or judges' salaries, are on average going to be in line with overall trends in the federal discretionary budget as a whole. Thus, long-term stability in R&D spending as a percentage of the whole budget is what we should expect to see exactly because of the decentralized essence of science policy.

Of course this reality means that federal support for S&T is subject to the same political processes and indignities as other federal discretionary programs. Although such a notion may offend the common claims of privilege made on behalf of publicly funded science, it also offers evidence of a durable embeddedness of S&T

in the political process as a whole, an embeddedness that has offered and will probably continue to offer significant protection against major disturbances in overall funding commitments for R&D activities. For this reason, predictions of impending catastrophe for research budgets (for example, in the mid-1990s, many science policy leaders believed that cuts of up to 20% in R&D were all but inevitable) have not come true. On the other hand, in periods of particular stress on the discretionary budget (and we are now in such a period) R&D faces the same budgetary pressures as other crucial areas of government budgetary responsibility, from managing national parks and supporting diplomatic missions to providing nutritional programs for poor infants and mothers or monitoring the safety of the nation's food supply.

Thus, any time of famine (or feast) for public civilian R&D funding as a whole will be a time of famine (or feast) for most other nonmilitary government programs subject to the annual budgeting process. Any argument that R&D deserves special protection from budgetary pressures is implicitly an argument that other programs are less deserving of protection. One sure way for R&D advocates to threaten the considerable stability of research funding in the budget would be to begin to target other, non-R&D programs as somehow less deserving of support. However compelling such arguments might seem to those who recognize the importance of a robust national investment in R&D, they will also be a provocation to those who are similarly compelled by competing priorities.

It will not have escaped the alert reader's notice that the 1960s do not fit into the story I have been telling so far. Civilian R&D funding relative to discretionary spending increased markedly in the early 1960s, peaked in 1965, and then declined to the levels that were to characterize the next 30 years. This excursion can be explained in one word: Apollo.

In the wake of Sputnik and at the height of the Cold War, President Kennedy's decision to send people to the Moon represents by far the most notable exception to the highly stable, buffered system that characterizes recent public funding for R&D: NASA's budget increased about 15-fold between 1960 and 1966. At the apogee of Apollo spending in 1966, nondefense R&D accounted for 25% of nondefense discretionary expenditures, but if you subtract the NASA component, the R&D investment falls to only about 6%. The perturbation was driven by external geopolitical forces; this was not the internal logic of scientific opportunity making itself felt

Figure 1: Non-defense R&D as Percent of Federal

Non-defense Discretionary Spending, FY 1962-2007

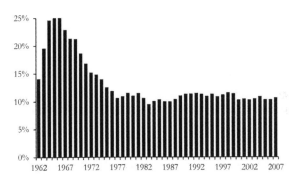

Source: OMB Historical Tables, Fiscal Year 2007

Apollo aside, the stability of the government's commitment to R&D as a proportion of its entire portfolio of discretionary activities also represents a commitment to growth. In 1960, before the Apollo ramp-up, nondefense R&D made up about 10% of nondefense discretionary spending, a level to which it returned after Apollo. Meanwhile, from 1962 (the first year for which reliable comparable date are available) to 2006, total nondefense inflation-adjusted R&D expenditures (in FY 2000 constant dollars) rose 335%, a rate of increase that closely mirrors general budgetary growth as a whole, rather than some natural rate of expansion of the knowledge-producing enterprise.

Of course these macro-scale trends conceal internal variations. NASA's rapid budgetary ascension was followed by a more gradual decay curve, but even at its 1974 post-Apollo perigee, NASA's budget of $3.2 billion (in current dollars) exceeded that of any other civilian R&D agency. NIH did not catch up to NASA until 1983, and did not leave it in the dust until the doubling began in 1998 (today, NIH's budget is almost 2.5 times NASA's). In 1977, in the wake of the Arab oil embargos, President Carter consolidated several agencies and programs to create the Department of Energy (DOE), with budgets of NASA-like magnitude. DOE's fortunes declined by almost 30% in terms of spending power under President Reagan, and today, after accounting for inflation, its budget is less than it was at its inception. NSF, whose importance for supporting university research belies its relatively modest share of overall nondefense R&D, has experienced budget increases in all but 2 of the past 42 years.

By remaining more or less aloof from focused political attention, NSF has avoided the volatility of DOE and NASA and, like the old blue-chip stocks, has yielded persistent if unspectacular growth. Its 2006 level of $5.5 billion ($4.2 billion for research) is still considerably less than that of NASA ($11.3 billion), DOE ($8.6 billion), or NIH ($28.4 billion). Overall, while civilian R&D as a whole is under the grip of a sort of budgetary lock-in due to larger political forces, the internal texture of the R&D budget is continually rewoven by dynamic political processes.

The stability of the federal commitment to S&T is matched, indeed exceeded, by a particular commitment to basic research. One of the most persistent myths of science policy is that government support for basic research is soft and has eroded over time. Indeed, the vulnerability of basic research to the vulgarities of politics has been an article of faith among science advocates at least since World War II, when Vannevar Bush, chief architect of NSF, explained in the classic report *Science, the Endless Frontier* that both government and industry were naturally inclined to favor investments in applied research over basic. Yet federal basic research investments (nondefense and defense) over the past 40 or so years have risen more quickly than R&D budgets as a whole. In 1962, the government investment in basic research was about 60% that of applied research; basic and applied reached parity in the late 1970s; and in recent years, driven especially by the NIH doubling, basic has exceeded applied by as much as 40%.

Despite the warnings of Vannevar Bush and subsequent science advocates, the political case for basic research is both strong and ideologically ecumenical. Unlike applied R&D, basic research appeals to the political left as an exemplar of the free expression of the human intellect, to the political right as an unambiguously appropriate area of government intervention because of the failure of the market to provide adequate incentives for private-sector investment, and to centrists as an important component of the government's role in stimulating high-technology innovation. In 1994, when Democrats lost their grip on Congress to a Republican majority bent on budget cutting, many of my scientist friends went into a panic, certain that academic basic research would be on the chopping block. But the value of federal investments in basic research was one thing that President Bill Clinton and House Speaker Newt Gingrich could agree on, and basic science fared well—better than it had under the Democrats in the first two years of the Clinton regime. Moreover, the general public, so often characterized as scientifically illiterate by politically illiterate scientists, has for decades shown very strong support for basic research in public opinion surveys.

Research that is "basic" is not necessarily irrelevant. Not only may scientists be curious about problems of abiding practical interest, but scientists may pursue questions that to them are of purely intellectual interest but to research administrators and policymakers are part of a more strategic effort to advance a particular mission. Economists such as Nathan Rosenberg and Richard Nelson and historians such as Stuart Leslie have demonstrated that basic research agendas have throughout the post–World War II era been strongly tied to the priorities of the private sector and national defense. Even arcane fields such as subatomic particle physics were justified during the Cold War in part because they were the training grounds for the nation's next generation of weapons scientists. The great majority of federal research categorized as basic is funded as part of larger agency missions, NIH being the most obvious example today, and the Department of Defense in earlier decades. NSF is the most conspicuous exception, although this, too, has been changing as NSF priorities increasingly focus on real-world priorities ranging from climate change to nanotechnology. Although such realities take the gloss off notions of scientific purity, they help to explain why, year after year, 535 members of Congress, most of whom have little if any deep knowledge of science, continue to treat basic science with a level of consideration equal to that of new post offices, interstate cloverleafs, and agricultural price subsidies.

The budgetary picture for R&D is not just about public investments, of course. When public and private support for R&D are considered together, evidence of consistent growth is even more pronounced, with inflation-adjusted expenditures rising from $71 billion in 1962 to $270 billion in 2004 (in FY 2000 dollars). Most of this growth has come in the private sector; indeed, this period shows a progressive decline in the ratio of federal to private funding for R&D. In 1962, the government funded twice as much R&D as did private industry. The continued growth of the high-technology economy led to increasing private-sector investment in R&D, and by 1980 the share of R&D funded by industry slightly exceeded the public share. By 2004, industrial R&D funding was more than twice that of government. From a simple market failure perspective, this trend represents a tremendous success: The private sector is taking on an increasing share of the knowledge-creation burden of society as the government investment brings increasing long-term economic returns.

Many good reasons exist for investing in science beyond just compensating for market failure. I want to emphasize that the continual attention paid to the amount of investment in R&D, whether public or private, military or civilian, tends to come at the expense of attention to these other reasons. One can easily imagine a variety of very different R&D portfolios for given levels of investment. Presumably each of those portfolios would contribute to very different sets of social outcomes. One could even imagine a large R&D investment portfolio organized in a way that contributes less to public well-being than a different, smaller portfolio.

The real world provides a way to begin thinking about such issues. R&D policies, and the resulting structures of national R&D enterprises, do vary

19

significantly from nation to nation. For example, among nations that invest substantially in R&D, industrial funding ranges from 75% of total R&D in Japan, to 63% in France, to 51% in Canada. Within the public investment sphere there are enormous differences in priorities among nations. The most obvious indicator of this diversity is biomedical science, which in the United States commands almost 50% of the total federal nondefense R&D budget, compared to 4% in Japan and Germany, 6% in France, and 20% in the United Kingdom. Similarly, Japan devotes about 20% of its civilian R&D to energy, whereas the United States spends about 3%, Germany 4%, France around 7%, and the United Kingdom 1%. The United States spends about 20% of its civilian R&D on space, France 10%, Germany 5%, and so on.

Such numbers don't tell the whole story, because European nations and Japan distribute large chunks of their federal R&D dollars in the form of block grants to universities, which then have discretion in allocating among various fields. Both within and between nations there is an enormous diversity of policy models used to determine R&D priorities, to translate those priorities into actual expenditures, and to apply those expenditures to science. In some ways the United States, with its Balkanized budgetary authority, is more decentralized and more diverse than most other affluent nations. At the same time, in the United States there is a tighter linkage between specific agency missions and funding allocation to research performers than in many other R&D-intensive nations. And of course the decision processes for the disbursement of funds in universities and national laboratories (as well as the relative roles of different types of R&D-performing institutions) vary greatly from nation to nation and within nations.

The role of peer review, smart managers, earmarks, block grants, and equity policies (for example, NSF's Experimental Program to Stimulate Competitive Research and German efforts to support institutions in the east) are all highly variable and reflect different institutional and national histories, politics, and cultures. Human resources are also variable, with the proportion of scientists and engineers at over 9 per 1,000 in Japan, 8 in the United States, 6 in France and Germany, and 5 in the United Kingdom. And of course the role of public R&D in "industrial policies" has varied greatly over time within the United States alone and varies greatly between nations and between sectors. The 1991 Office of Technology Assessment report Federally Funded Research: Decisions for a Decade summed up the situation: "While there may be certain universality in science, this does not carry over to science policy."

These details are the very nuts and bolts of what science and technology policy is supposed to be all about, and much emotion and energy are invested in promoting policies that favor one approach, priority, or program over another. But given the great diversity in science policies both within and among affluent nations, and given the relative similarity of the macroeconomic and socioeconomic profiles of these same nations, I can see no reason to believe that there is a strong linkage between specific national science policies and general national-scale socioeconomic characteristics. Of course, the United States has a strong aerospace industry and a strong pharmaceutical industry in part because of R&D policy priorities during the past 50 years, but it still has a 20% pretax poverty rate, average life expectancy in the mid-70s, gross domestic product per capita above $30,000, a Human Development Index above 0.9, and so on, just like other affluent nations with very different approaches to investing in R&D. (The United States also has famously mediocre public health indicators despite its gargantuan investment in biomedical research.) So, although federally sponsored S&T are obviously causal contributors to public welfare and although S&T policies of some sort are necessary to ensure such contributions in the future, there is little reason to imagine that, at the macro policy level, particular policy models and choices make much of a difference to broad socioeconomic outcomes. There seems to be a diverse range of options that work more or less equally well, and this diversity may itself be a component of success.

From this perspective, the machinations of science policy — the constant stream of conferences, reports, and newspaper op-eds; the dozens of committees and working groups; the lobbying and legislation; the hyperbole and anxiety — are best viewed as metabolic byproducts of a struggle for influence and funding among various political actors such as members of Congress, executive-branch administrators, corporate lobbyists, college presidents, and practicing scientists. The significance of this struggle is largely political and internal to the R&D enterprise; it is not a debate over the future of the nation, despite continual grandiose claims to the contrary. We are mostly engaged in science politics, not science policy. Or, to adopt the perspective of Thomas Kuhn, this is normal science policy, science policy that reinforces the status quo.

Publicly funded research is justified on the basis of promised contributions to desired social outcomes: to "increase quality and years of healthy life [and] eliminate health disparities" (U.S. Department of Health and Human Services), to "conserve and manage wisely the Nation's coastal and marine resources" (National Oceanic and Atmospheric Administration), or to ensure "a safe and affordable food supply" (U.S. Department of Agriculture). What is the capacity of a particular science policy decision to advance a given desirable outcome?

This should be the most fundamental science policy question, because if one cannot answer it, then one cannot know whether any particular policy is likely to be more or less effective than any alternative policy. And if one cannot choose among alternative policies in terms of what they may achieve, then policy preferences are revealed as nothing more than expressions of parochial values and interests. This turns out to be a perfectly good approach to ensuring that R&D investments are treated as well as other public investments. But it does not tell us whether different investment choices would yield better (or worse) outcomes.

At the level of "national" R&D policies, it appears that different approaches yield more or less similar outcomes. But when trying to connect R&D to particular desired outcomes, policy choices obviously can matter greatly. The doubling of the NIH budget is a case in point. This doubling occurred without any national dialogue about what it might achieve or about alternative paths toward better national and global health. In part because of the close coupling between NIH research agendas and biomedical industry priorities, high-technology intervention (often at high cost, as well) has been adopted as the national strategy for improved health, and no serious consideration was given to alternative health investment strategies that might be equally effective in contributing to public well-being but less likely to contribute to significant corporate profitability.

Even further from the debate was the question of whether biomedical research was the area of science that could yield the most public value from a rapid increase in investment, rather than, say, energy R&D. One might compellingly have made the case that the nation's health challenges are far less of an immediate threat to well-being than its dependence on fossil fuels imported in large part from other nations. That there are no mechanisms or forums to explore these types of tensions between alternative approaches to a particular goal, such as better health, or between competing goals, such as better health or better energy, is precisely the problem.

The internal political dynamics of science budgeting help to explain why R&D policy discussions are dominated by concerns about "how much" and avoid like the plague serious questions about "what for." But I'm suggesting here that the "how much" obsession may paradoxically reduce the potential contribution of R&D to well-being, because "more" and "better" are simply not the same things. For example, "how much" carries with it a key, but unstated, assumption: that everyone is made better off by investments in science. If the benefits of science are broadly and equitably distributed, then "how much science can we afford?" is a reasonable central question for science policy, especially given the decentralization of the policy process.

Whatever the priorities may be, we can expect that all will benefit. But if the positive and negative effects of science are unevenly distributed, the primacy of "how much" becomes more difficult to justify from a perspective of good governance and good government. And the positive and negative effects of science are indeed unevenly distributed. Besides, given what we know about such problems as unequal access to health care and the disproportionate exposure of poor people to environmental hazards, my colleague Edward Woodhouse and I have suggested a hypothesis that seems worthy of careful testing: New scientific and technological capacities introduced into a highly stratified society will tend disproportionately to benefit the affluent and powerful.

It is not very difficult to imagine the types of questions that might help to inform a transition from science policy based on "how much" to science policy based on "what for," though it is certainly the case that such questions may be rather unwelcome in national R&D policy discussions and that even partial answers will not always be available, at least at first. But here are 10 questions that, if made explicit in science policy discussions, could help with the transition:

1. What are the values that motivate a particular science policy?

2. Who holds those values?

3. What are the actual goals that the policy is trying to achieve?

4. What are the social and institutional settings in which the R&D information or products will be used?

5. What are the reasons to expect that those are settings for effectively translating the results of R&D into the goals that justify the policy?

6. Who is most likely to benefit from the translation of the research results into social outcomes?

7. Who is unlikely to benefit?

8. What alternative approaches (through either other lines of research or non-research activities) are available for pursuing such goals?

9. Who might be more likely to benefit from choosing alternative approaches?

10. Who might be less likely to benefit?

Addressing these questions does not require impossible predictions of either the direction of scientific advance or the complex interactions between science and society. It does require that unstated agendas and assumptions, diverse perspectives, and the lessons of past experiences be part of the discussion. The questions are as appropriate for academic scholarship as they are for congressional hearings or media inquiries. Taking them seriously would be a step toward a science policy that mattered.

3

ENDLESS FRONTIERS, LIMITS OF KNOWLEDGE, AND THE FUTURE OF THE HUMAN CONDITION

Michael Crow

During the past fifty years, many of us have come to terms with an increasing realization that there may be a limit to what we as a species can plan or accomplish. The U.S. failure to protect against and respond to Hurricane Katrina in the summer of 2005 and the apparent futility of the plan to democratize and modernize Iraq provide particularly painful evidence that we seem to be operating beyond our ability to plan and implement effectively, or even to identify conditions where action is needed and can succeed.

Our disappointing performances, as a human species, in "flooded" New Orleans and "liberated" Iraq might be more encouraging if these two geographies presented us the most complex problems we need to address, but they are child's play compared to the looming problems of global terrorism, climate change, or possible ecosystem collapse; problems that are not only maddeningly complex but also potentially inconceivably destructive.

Our current approach to framing problems can be traced back to the 1972 publication of the Club of Rome's The Limits to Growth, which posed the still-unanswered question: How much population growth and development, how much modification of natural systems, how much resource extraction and consumption, and how much waste generation can Earth sustain without provoking regional or even global catastrophe? Since that time, the way we think about human activity and the environment and the way we translate this thinking into our science policy and subsequent R&D, public debate, and political action have been framed by the idea of external limits — defining them, measuring them, seeking to overcome them, denying their existence, or insisting that they have already been exceeded.

For technological optimists these limits are ever-receding, perhaps even nonexistent, as science-based technologies allow progressive increases in productivity and efficiency that allow the billion and a half people living in industrialized and industrializing nations today to achieve a standard of living that was unimaginable at the beginning of the 20th century. For the pessimists, there is global climate change, the ozone hole, air and water pollution, overpopulation, natural and human-caused environmental disasters, widespread hunger and poverty, rampant extinction of species, exhaustion of natural resources, and destruction of ecosystems. In the face of these conflicting perceptions, it makes no sense to try to use external limits as a foundation for inquiry and action on the future of humans and the planet. It is time to look elsewhere.

All sides in the limits-to-growth debate would

probably agree on the following two observations: First, the dynamic, interactive system of complex biogeochemical cycles that constitutes Earth's surface environment is falling increasingly under the influence of a single dominant life form: us. Second, this life form, notable for its ability to learn, reason, innovate, communicate, plan, predict, and organize its activities, nonetheless exhibits serious limitations in all these same areas.

During the past 150 years, scientific and technological innovation has facilitated enormous growth: The population of Earth has increased approximately six fold, the average life span of those living in the industrialized nations has doubled, agricultural productivity has increased by a factor of five, the size of the U.S. economy alone has increased more than 200-fold, the number of U.S. scientists has increased by more than 17 times, and the volume of globally retrievable information stored in analog and digital form has expanded by incalculable orders of magnitude. At the same time, 20% of the planet's bird species have been driven into extinction, 50% of all freshwater runoff has come to be consumed, 70,000 synthetic chemicals have been introduced into the environment, the sediment load of rivers has increased fivefold, and more than two-thirds of the major marine fisheries on the planet have been fully exploited or depleted.

As Joel Cohen has brilliantly illustrated in his book *How Many People Can the Earth Support?*, there are many possible futures available to us. The only certainty is that present trajectories of growth cannot, and therefore will not, be maintained indefinitely. (Thomas Malthus got this point right more than 200 years ago. He simply failed to appreciate the productivity gains that science and technology could deliver.) The central question that faces us is whether we will be able to position ourselves to choose wisely among alternative future trajectories or will simply blunder onward. The markets will indeed adjust to the eventual depletion of fossil-fuel reserves, for example, but will likely be too shortsighted to prevent global economic disruption on an unprecedented scale, a situation that could even lead to global war.

If we continue to define our problem as external to ourselves—as limits imposed by nature and the environment—then we consign ourselves to a future of blundering.

The limits that matter are internal. They are the limits on our collective ability to acquire, integrate, and apply knowledge.

Although it is difficult to isolate these limits neatly from one another, it is helpful to separate them into six large categories: limits of the individual, of sociobiology, of socioeconomics, of technology, of knowledge, and of philosophy. Although these might at first seem to be insurmountable shortcomings, I believe that our best hope for finding our place in nature and on the planet resides in embracing our limits and recognizing them as explicit design criteria for moving forward with our knowledge production and organization. I see potential for progress in each.

Individual limits. We all operate out of self-interest, which is entirely rational. Community spirit and altruism may be motivating factors, but given that we cannot know the effects of our individual actions on the larger systems in which we are enmeshed, the only reasonable alternative is for each of us to pursue our conception, however imperfect, of our own interests. Yet as social systems grow more and more complex and as they impinge more and more on natural systems, our individual vision inevitably captures less and less of the big picture. Our only option is to accept the limits of individual rationality and to take them into account in formulating public policy and collective action.

Sociobiological limits. During the course of our development, humanity's special capabilities in areas such as toolmaking, language, self-awareness, and abstract thought have rendered us extraordinarily fit to engage in the competitive business of individual and species survival. We compete among ourselves at every organizational level and with other species in virtually every ecological niche. Cooperation, therefore, most often occurs at one level (a tribe or a nation, for example) in order to compete at a higher level (a war between tribes or nations). But at the highest levels—the behavior of an entire species competing with or dominating billions of other species—we have run out of reasons to cooperate or structures to foster effective cooperation. We need to consciously search for ways to transcend our sociobiological limits on cooperation.

Socioeconomic limits. We have done our best to make a virtue out of our individual and sociobiological limits through market economics and democratic politics. Yet we are unable to integrate the long-term consequences of our competition-based society into our planning processes. Our competitive nature values the individual over the group, but the aggregation of individual actions constantly surprises us. Despite our best intentions, our actions are consistent with a global economy predicated on the expectation of continued growth and development derived from ever-increasing resource exploitation. Thus, for example, we all climb into our cars in the morning thinking only that this is the most convenient way to get to work. We are not deliberately choosing to waste time in traffic jams, exacerbate the trade deficit, and pump greenhouse gases into the atmosphere. But we find it extraordinarily difficult to anticipate or accurately account for the costs and risks incurred over the long term by such group behavior. Indeed, those costs and risks vary wildly from individual to individual and from group to group. An example of this is the cost/benefit calculation that must have been made regarding New Orleans, where the probability of catastrophic flooding is low and the cost of protecting the city is high. At every level of the political system, the individual perspective outweighed the collective, with the result that adequate protection for the whole community lost out. Because of these complexities, efforts to advance the long-term interests of the whole by controlling the short-term behavior of the individual are doomed to failure, which is one of the lessons of the global collapse of communism.

Technological limits. To evade the behavioral limits of biology and economics, we have turned to technology. Indeed, technology, harnessed to the marketplace, has allowed industrialized societies to achieve amazingly high standards of living. In doing so, however, we have put our future into the hands of the lowest bidder. Cheap oil and coal, for example, ensure our continued dependence on the internal combustion engine and the coal-burning power plant. The problem we face is not a shortage of polluting hydrocarbon fuels, but an excess. History shows that we will develop increasingly efficient energy technologies but that gains in efficiency will be more than offset by the increased consumption that accompanies economic growth. The increased efficiency and cleanliness of today's cars when compared with those built in 1980 are an example. Technology has allowed us to pollute less per mile of driving, but pollution has declined little because we drive so many more miles. Too often we choose technologies that save us from today's predicament but add to the problems of tomorrow.

Knowledge limits. There is absolutely no a priori reason to expect that what we can know is what we most need to know. Science uses disciplinary organization to recognize and focus on questions that can be answered. Disciplines, in turn, are separated by methodology, terminology, sociology, and disparate bodies of fact that resist synthesis. Although disciplinary specialization has been the key to scientific success, such specialization simultaneously takes us away from any knowledge of the whole. Today the whole encompasses six billion people with the collective capability of altering the biogeochemical cycles on which we depend for our survival. Can science generate the knowledge necessary to govern the world that science has made? Do we even know what such knowledge might be? Producing 70,000 synthetic chemicals is easy compared to the challenge of understanding and dealing with their effects. Despite the billions we have spent studying our interference with the planet's biogeochemical cycles, we really do not have a clue about what the long-term result will be. And we have even less knowledge about how to organize and govern ourselves to confront this challenge. The intrinsic difficulties of creating a trans-disciplinary synthesis are compounded dramatically by a dangerous scientific and technological illiteracy among senior policy-makers and elected officials. An ironic effect of technology-created wealth is the growth of an affluent class that prizes individualism over civic engagement and that feels insulated from the need to understand and confront complex technology-related social issues.

Philosophical limits. The scientific and philosophical intellectuals of "the academy" remain focused on the relatively simple question of understanding nature. The much more complicated and challenging—and meaningful—quest is to understand nature with a purpose, with an objective, with an end. What is the purpose of our effort to understand nature: to learn how to live in harmony with nature or to exploit it more efficiently? For thousands of years, philosophical inquiry has been guided by such fundamental questions as "Why are we here?" and "How should we behave?" Such questions were difficult enough to confront meaningfully when our communities were small, our mobility limited, and our impact restricted. In today's hyperkinetic world, how can we possibly hope to find meaning? The literal answers provided by science amount to mockery: We are here because an expanding cloud of gas some 15 billion years ago eventually led to the accretion of planets, the formation of primordial nucleotides and amino acids, the evolution of complex organisms, the growth of complex social structures in primates, and the dramatic expansion of cognitive and analytical capabilities made possible by the rapid evolution of neocortical brain structures. Such explanation is entirely insufficient to promote the commonality of purpose necessary for planetary stewardship. We lack a unified or unifiable metaphysical basis for action, just when we need it most.

I list these limits—which no doubt could be parsed and defined in many different ways—not to bemoan them, but to acknowledge the boundary conditions that we face in learning how to manage our accelerating impact on Earth. How can we create knowledge and foster institutions that are sensitive to these boundary conditions? This is a sensitivity that we have hardly begun to develop and that will not be found in any of compartmentalized traditional disciplines that we nurture so earnestly. Not only do we perpetuate traditional disciplines, we assign inordinate significance to distinctions in a strict hierarchy: disciplines trump other disciplines based on their quantitative capacities. The academy remains unwilling to fully embrace the multiple ways of thinking, the different disciplinary cultures, orientations, and approaches to solving problems that have arisen through hundreds if not thousands of years of intellectual evolution. Our science remains culturally biased and isolated: Western science is derivative of a philosophical model of domination and the manipulation of nature, as opposed to the acceptance of natural systems and dynamics.

The problems that we face are not hierarchical, nor do they fall within strict disciplinary categories. They require multiple approaches and an integration of disciplines; we cannot expect biologists alone to solve the problem of the loss of biodiversity. Because each academic discipline has a Darwinian focus on its own survival, none has the impetus or the capacity to develop a formal language to make itself comprehensible to other disciplines. We have not developed the means for chemists to talk to political scientists, and for political scientists to talk to earth scientists, and for earth scientists to talk to engineers. The debate must engage a broad community of disciplines, and not just the expertise found within the universities but also the wisdom and expertise developed in commerce, industry, and government.

We need new ways to conceive of the pursuit of knowledge and innovation, to understand and build political institutions, to endow philosophy with meaning for people other than philosophers. We trumpet the onset of the "knowledge society," but we might be much better off if we accepted that, when it comes to our relations with nature, we are still pretty much an "ignorance society." Our situation is reminiscent of Sherman McCoy, the protagonist of Tom Wolfe's *Bonfire of the Vanities*, who fancies himself a "Master of the Universe" just as his life is taken over by events far beyond his control. We have the illusion of understanding and are not humbled by the fact that we do not understand. We refuse even to consider the possibility. Hubris, exemplified in the demands we make on science, is a major obstacle to coming to grips with our situation. We are obsessed with trying to predict, manage, and control nature, and consequently we pour immense intellectual and fiscal resources into huge research programs—from the Human Genome Project to the U.S. Global Change Research Program—aimed at this unattainable goal. On the other hand, we devote little effort to the apparently modest yet absolutely essential question of how, given our unavoidable limits, we can manage to live in harmony with the world that we have inherited and are continually remaking.

Concepts such as sustainability, biodesign, adaptive management, industrial ecology, and intergenerational equity—new principles for organizing knowledge production and application—offer hints of an intellectual and philosophical framework for creating and using knowledge appropriate to our inherent limits. Sustainability is a concept as potentially rich as justice, liberty, and equality for guiding inquiry, discourse, and action. Biodesign seeks to mimic and harness natural processes to confront challenges in medicine, agriculture, environmental management, and national security. Adaptive management acknowledges the limits of acquiring predictive understanding of complex systems, and although the prospect of their control is illusory, the genesis of increasingly sophisticated data sets should impart increasing "predictability" to the bandwidth in which systems may behave. Industrial ecology responds to our tendency to organize and innovate competitively, and looks to natural systems for a model of innovation that can enhance competitiveness while reducing our footprint on the planet. Intergenerational equity seeks to apply core societal values such as justice and liberty across boundaries of time as well as space. Of course, we will need many other new ways to think about and organize our actions, but these few indicate a beginning.

Common to all such approaches is the idea that more flexibility, resilience, and responsiveness must be built into all institutions and organizations—in academia, the private sector, and government—because society will never be able to control the large-scale consequences of its actions. In today's "ignorance society," we must define some measure of rationality and recognize that the only way to reduce uncertainty about the future is to take action and carefully observe the outcomes. We must establish threshold criteria for, or at least attempt to define, the range of potential scenarios for which some degree of planning either to promote or obstruct a given outcome should be contemplated. The latter is the more difficult, particularly if a major risk or disaster begins to emerge. Yet we should not succumb to the paralysis of the "precautionary principle," which saps innovation and risk-taking. The more institutional and organizational innovation we conduct, the better the chances that we will learn how to deal with the implications of our own limits.

The ideological and institutional struggle between communism and market democracy can be viewed as one such set of competing innovations, albeit poorly planned and exceedingly costly. A key result of this innovation competition is the certain knowledge that rational self-interest cannot be successfully suppressed indefinitely and that legal systems that foster dissent and freedom of choice provide a fertile culture for innovation. We now urgently need to conceptualize a new series of innovations, at much lower cost and shorter run-time, to push this result further and apply it to the problem of ensuring that our global society can continue to be sustained by the web of biogeochemical cycles that makes life possible in the first place.

4

DEMOCRATIZING SCIENCE:
ENDS, MEANS, OUTCOMES

David Guston

Science is deeply political, and always will be. Asking whether science is politicized — or can be de-politicized — distracts us from asking. "Who benefits and loses from which forms of politicization?" and "What are the appropriate institutional channels for political discourse, influence, and action in science?" Arguing over whether science is politicized neglects the more critical question: "Is science democratized?"

Democratizing science does not mean settling questions about nature by plebiscite, any more than democratizing politics means setting the prime rate by referendum. What democratization does mean, in science as elsewhere, is creating institutions and practices that fully incorporate principles of accessibility, transparency, and accountability. It means considering the societal outcomes of research at least as attentively as the scientific and technological outputs. It means insisting that in addition to being rigorous, science can be popular, relevant, and participatory.

These conceptions of democratization are neither new nor, when applied to science, idiosyncratic. They have appeared in discussions about science at critical historical junctures. For example, the Allison Commission, a congressional inquiry into the management of federal science in the 1880s, established the principle that even the emerging "pure science" would, when publicly financed, be subject to norms of transparency and accountability, despite John Wesley Powell's protestations. After World War II, the creation of the National Science Foundation (NSF) hinged on establishing a politically accountable governing structure. These concerns exist at the heart of arguments made by theorists such as Columbia University philosopher Philip Kitcher, who describes the accessible and participatory ideal of "well-ordered science" in his *Science, Truth, and Democracy*. They likewise exist in many current science agencies and programs, but there they often fly under the radar of higher-profile issues or have been institutionalized in ways that undermine their intent. They do not exist, however, as an agenda for democratizing science. Below, I attempt to construct such an agenda: a slightly elaborated itemization of ways to democratize both policy for science and science in policy.

In the past, critics of elite science attempted to democratize policy for science by expanding the array of fields that the federal government supported, as Sen. Harley Kilgore attempted to do with the social sciences in the early debate over NSF, or by creating programs that were explicitly focused on societal needs, as Rep. Emilio Daddario did with NSF's Research Applied to National Needs. These approaches were problematic because public priorities are just as easily hijacked by disciplinary priorities in the social sciences as in the natural sciences. Moreover, at a basic research institution such as NSF, applied research may be either too small to have great influence on the larger society or just large enough to threaten the pure research mission. My agenda for democratizing policy for science takes a different tack by broadening access across the sciences and across the levels at which priorities are set.

First, engage user communities and lay citizens more fully in review of funding applications. Such "extended peer review" increases the presence of public priorities without mandating research programs or diluting quality. The National Institutes of Health (NIH) pioneered a modest form of extended peer review by including citizens on its grant advisory councils, but the councils' reviews of study sections' recommendations have a pro forma quality. The NIH Web site acknowledges that "the use of consumer representatives may be extremely helpful in the review of certain areas of research," but it still holds "it is often neither necessary nor appropriate to include consumer representatives in peer review."

A more thorough use of extended peer review occurs at the National Institute on Disability and Rehabilitative Research of the Department of Education, which seeks input from relevant disability communities in funding decisions and post-hoc review. Disciplinary research such as that supported by NSF would be less likely to benefit from such input, although priorities across areas of inquiry, such as climate research, would benefit from an understanding of what public decision-makers want and need to know. For the vast majority of mission-oriented public R&D spending, such participation is likely a better way to ensure the conduct of basic research in the service of public objectives, a goal sought by a diverse set of analysts, including Lewis Branscomb

41

and Gerald Holton ("Jeffersonian science"), Donald Stokes ("Pasteur's Quadrant"), and Rustum Roy ("purposive basic research"), not to mention policy-makers Sen. Barbara Mikulski ("strategic research") and the late Rep. George Brown ("science in service of society").

Second, increase support for community-initiated research at universities and other research institutions. National R&D priorities are driven by large private investments. Through changes in intellectual property, public investments have become increasingly oriented toward the private sector, even as private R&D spending has grown to twice the size of public R&D spending. "Science shops" — research groups at universities that take suggestions for topics from the local citizenry — offer the opportunity for community-relevant priorities to emerge from the bottom up. This research might include more applied topics that are unlikely to draw grant money, such as assessments of local environmental health conditions. It might also facilitate connections between research universities and local economic interests that are less dependent on intellectual property.

These connections would be akin to agricultural or manufacturing extension, and they could be funded in the same politically successful way. By allowing some of the priorities of the research enterprise to emerge more directly from local communities, science shops can help reinvigorate the concept of "public interest science," articulated in the 1960s by Joel Primack and Frank Von Hippel, and help set a research agenda that is not captive to large economic interests.

Third, restructure programs in the ethical, legal, and societal implications (ELSI) of research. If ELSI programs, such as those funded with the genome or nanotechnology initiatives, are to facilitate democratic politics and improve the societal impacts of knowledge-based innovation, they need to meet two criteria. First, they must extend into research areas that have not already been designated for billion-dollar public investments. Such a change would not only protect them from being swamped by the mere scale of technical activity but would also allow them to identify technical areas prospectively and have an influence on whether and how such large-scale public investments are made. Second, ELSI research must be more directly plugged back into the policy process. ELSI programs should include more technology assessment and "research on research," areas that can contribute to understanding the role of science and technology in broader political, economic, and cultural dynamics, but from which the federal government has pulled back intramural resources. ELSI programs should also have institutional connections to decision-makers, as the genome program initially did. In addition to setting aside three to five percent of the R&D megaprojects for ELSI work, the federal government should set aside a similar amount for all R&D programs over a certain size, perhaps $100 million, and should fund much-expanded research programs in the societal dynamics of science and technology through NSF.

Discussion of the democratization of science advice borders on the current controversy over politicization. Despite their recent political currency, issues of science advice will not attract media or move voters in the way that issues of guns and butter will, and thus the circuit of transparency and accountability will be incomplete. In earlier periods of reform, concerns about the politics and process of expert advice led to the Federal Advisory Committee Act, which mandates transparency in the actions of advisory committees and balance in their membership. A report by the Government Accountability Office (GAO) found that agencies need better guidance to implement the balance requirement, but more wide-ranging action is needed.

First, recreate an Office of Technology Assessment (OTA) to restore the policy-analytic balance between Congress and the Executive Branch in matters scientific and technological. Without competition from a co-equal branch, Executive-based science advice has a monopoly, and monopolies in the marketplace of ideas do not serve democracy. There have been multiple, behind-the-scenes efforts to reconstitute a congressional capacity for technology assessment, including projects at GAO. A positive finding from an independent evaluation of an initial pilot project encouraged Representatives Holt, Houghton, Barton, and Boehlert to draft a bill authorizing $30 million for an Office of Science and Technical Assessment (OSTA) in GAO. The bill specified that OSTA assessments would be publicly available, thus contributing to democratic politics as well as providing competition for Executive Branch expertise. Even if OSTA is authorized and funded, its influence would remain to be seen. But establishing OSTA would create, at least in part, a public deliberative space for science and policy that a modern democracy requires.

Second, enhance the transparency and accountability of expert deliberations through discussion and articulation of science policy rules. The decision rules for guiding how experts provide science advice require more scrutiny and better articulation. Even supposing that science advice were purely technical, any group of experts larger than one still needs a set of decision rules by which to settle disagreement. The character of such rules (for instance, linear and threshold models for assessing risk), is familiar in environmental policy. Such rules also include the admissibility of evidence, the definition of expertise and conflicts of interest, the burden and standards of proof, and the mechanisms for aggregating expert opinion.

A particular example of the last rule would be instituting recorded votes within expert advisory committees, rather than pursuing a vague consensus as most panels do. Committees of the National Toxicology Program make recommendations for the biennial Report on Carcinogens by recorded vote, and it seems salutary as it both specifies the relative level of agreement within the committee and creates a record that can be used to assess the objectivity and balance of a committee, thus providing information for a more democratic politics of expertise. A second example is the Supreme Court's *Daubert* decision, which describes considerations that trial judges should apply when deciding on the admissibility of expert testimony. Every venue of expert deliberation evaluates expertise implicitly or explicitly, yet the rules for such evaluations are rarely the focus of study, public discussion, or democratic choice.

Third, increase the opportunities for analysis, assessment, and advice-giving through the use of deliberative polling, citizens' panels, and other participatory mechanisms. Such "participatory technology assessment" circulates views among citizens and experts, promotes learning about both science and democracy, and generates novel perspectives for policy-makers.

These mechanisms are more familiar in European settings, where the Danish Board on Technology uses citizens' panels for public education and government advising, and the Netherlands Office of Technology Assessment develops other forms of public input. NSF has funded quasi-experiments in face-to-face and Internet-mediated citizens' panels, and the Nanotechnology Research and Development Act endorses the use of such panels, among other outreach techniques, to inform the National Nanotechnology Initiative (an arrangement that also connects ELSI to policy).

At Arizona State University, the Consortium for Science, Policy, and Outcomes is implementing a research agenda called "real-time technology assessment" that combines traditional technology assessment with historical, informational, and participatory approaches in an effort to incorporate intelligent feedback into knowledge-based innovation. One could imagine building the capacity to foster exchanges among experts, citizens, and civic organizations at all major research universities — not to replace more technocratic methods, but as a necessary complement for a system of democratic science advice, analysis, and assessment.

Some readers will surely find this agenda not nearly far-reaching enough to democratize science. Others will just as surely think it threatens the autonomy and integrity of science. And there are most certainly grander ways of perfecting our democracy that, although not directly dealing with science, would transform it as well. Such betwixt and between may be uncomfortable rhetorically, but I think it wise politically.

Science and democracy have both been around for a long time without being perfected, and my agenda will not complete the task. These incremental steps, aimed at further implementing broadly recognized values of accessibility, transparency, and accountability, will admittedly not democratize science immediately and thoroughly. Neither will they condemn it to populist mediocrity. What pursuing this agenda might do, however, is foster the intellectual and political conditions for a relatively more democratic science to flourish within the current wanting environment. Discussing this agenda may, at the very least, shift the focus from sterile argument over politicizing science to deliberation about democratizing science.

5

SCIENCE POLICY:
BEYOND THE SOCIAL CONTRACT

David Guston

A widely held tenet among policy scholars maintains that the way people talk about a policy influences how they and others conceive of policy problems and options. In contemporary political lingo, the way you talk the talk influences the way you walk the walk.

Pedestrian as this principle may seem, policy communities are rarely capable of reflexive examinations of their rhetoric to see if the words used, and the ideas represented, help or hinder the resolution of policy conflict. In the science policy community, the rhetoric of the "social contract for science" deserves such examination. Upon scrutiny, the social contract for science reveals important truths about science policy. It evokes the voluntary but mutual responsibilities between government and science, the production of the public good of basic research, and the investment in future prosperity that is research.

But continued reliance on the concept of the social

contract, and especially calls for its renewal or re-articulation, are fundamentally unsound. Based on a misapprehension of the recent history of science policy and on a failed model of the interaction between politics and science, such evocations insist on a pious rededication of the polity to science, a numbing re-articulation of the rationale for the public support of research, or an obscurantist re-systemization of research nomenclature. Their effect is to distract from a new science policy, what I call "collaborative assurance," that has been implemented for 30 years, albeit in a haphazard way.

One cannot travel the science policy corridors of Washington, D.C., or read the pages of science policy journals, without stumbling across the "social contract" for science. The social contract for science is part of the science policy scripture, including work by Harvey Brooks, Bruce Smith, Donald Stokes, and others. Its domain is catholic, even global: the World Conference on Science, co-organized by the United Nations Educational, Scientific, and Cultural Organization and the International Council for Science, has called for a "new social contract" that would update terms for society's support for science and science's reciprocal responsibilities to society.

In my book *Between Politics and Science*, I unearth a more complete genealogy of the social contract for science, pinpoint its demise three decades ago, and discuss the policies created in its wake. I find its origin in two affiliated concepts: the actual contracts and grants that science policy scholar Don K. Price placed at the center of his understanding of the "new kind of federalism" in the relationship between government and science; and a social contract for scientists, a relationship among professionals that the sociologist Harriet Zuckerman described as critical to the maintenance of norms of conduct among scientists. Either or both of these concepts could have evolved into the social contract for science.

Most observers associate the social contract for science with Vannevar Bush's report Science, The Endless Frontier, published at the end of World War II. But Bush makes no mention in his report of such an idea and neither does John Steelman in his Science and Public Policy five years later. Yet commonalities between the two, despite their partisan differences, point toward a tacit understanding of four essential elements of postwar science policy: the unique partnership between the federal government and universities for the support of basic research; the integrity of scientists as the recipients of federal largesse; the easy translation of research results into economic and other benefits, and the institutional and conceptual separation between politics and science.

These elements are essential because they outline the postwar solution to the core analytical issue of science policy: the problem of delegation. Difficulties arise from the simple fact that researchers know more about what they are doing than do their patrons. How then do the patrons assure themselves that the task has been effectively and efficiently completed, and how do the researchers provide this assurance? The implications of patronage have a long history: from Galileo's naming the Medician stars after his patron; to John Wesley Powell's assertions in the 1880s that scientists, as "radical democrats," are entitled to unfettered federal patronage; to research agencies' attempts to meet the requirements of the Government Performance and Results Act of 1993.

How politics and science go about solving the problem of delegation has changed over time. The change from a solution based on trust to one based on "collaborative assurance" marks the end of the social contract for science.

The problem of delegation is described more formally by principal-agent theory, where the principal is the party making the delegation, and the agent is the party performing the delegated task. In federally funded research, the government is the principal and the scientific community the agent. One premise of principal-agent theory is an inequality or asymmetry of information: The agent knows more about performing the task than does the principal. This premise is not a controversial one, particularly for basic research. It is exacerbated by the near-monopoly that exists between government support and academic performance, which permits no clear market pricing for basic research or its relatively ill-defined outputs.

This asymmetry can lead to two specific problems (described with jargon borrowed from insurance theory): adverse selection, in which the principal lacks sufficient information to choose the best agent; and moral hazard, in which the principal lacks sufficient information about the agent's performance to prevent shirking or other misbehavior.

The textbook example of adverse selection is the challenge health insurers face. The people most interested in obtaining health insurance are those most likely to need it, and are thus most likely to cost more to insure. But their health problems are better known to the themselves than to the insurer. The textbook example of moral hazard is when the provision of fire insurance also provides an incentive for arson. Insurers attempt to reduce these asymmetries through expensive monitoring strategies, such as employing physicians to conduct medical examinations or investigators to examine suspicious fires. They also provide explicit incentives for behaviors to reduce the asymmetries, such as lower premiums for avoiding health risks such as smoking, or credits for installing sprinkler systems.

Both adverse selection and moral hazard operate in the public funding of basic research. The peer review system, in which the choice of agents is delegated to a portion of the pool of potential agents themselves, monitors the problem of adverse selection. Although earmarking diverts funds from it and critics question it as self-serving, peer review has been expanding its jurisdiction in the choice of agents beyond the National Science Foundation (NSF) and the National Institutes of Health (NIH). But immediately after World War II, there was no prominent consensus supporting the use of peer review to distribute federal research funds, and thus it was not part of any social contract for science that could have originated then.

Moreover, regardless of the mechanism for choice, the funding of research always confronts moral hazards that implicate the integrity and productivity of research. The asymmetry of information makes it difficult for the principal to ensure and for the agent to demonstrate that research is conducted with integrity and productivity. In Steelman's words: "The inevitable conclusion is that a great reliance must be placed upon the intelligence, initiative, and integrity of the scientific worker."

The social contract for science relied on the belief that self-regulation ensured the integrity of the delegation and that the linear model, which envisions inevitable progress from basic research to applied research to product and service development to social benefit, ensured its productivity. Unlike health or fire insurance providers, the federal government did not monitor or deploy expensive incentives to assure itself of the success of the delegation. Rather, it conceived a market-like model of science in which important outcomes were assumed to be automatic. In short, it trusted science to have integrity and be productive.

There were, of course, challenges to the laissez faire relation between politics and science, including conflicts over the loyalty of NIH- and NSF-funded scientists during the early 1950s, the accountability of NIH research in the late 1950s and early 1960s, the relevance of basic research to military and social needs in the late 1960s and early 1970s, and the threat of novel risks from genetic research in the 1970s. Some of these challenges led to modest deviations from the upward trajectory of research funding. But even issues that led to procedural changes in the administration of science, including the Recombinant DNA Advisory Committee, failed to alter the institutionalized assumption of automatic integrity and productivity.

Reliance on the automatic provision of integrity and productivity by the social contract for science began to break down, however, in the late 1970s and early 1980s. Well before the high-profile hearings conducted by Rep. John Dingell (D-Mich.) into allegations involving Nobel laureate David Baltimore, committees in the House and Senate scrutinized cases of scientific misconduct. The scientific community downplayed the issue. Philip Handler, then president of the National Academy of Sciences, testified that misconduct would never be a problem because the scientific community managed it in "an effective, democratic, and self-correcting mode."

To assist the community, Congress passed legislation directing applicant institutions to deal with misconduct through an assurance process for policies and procedures to handle allegations. But believing that public scrutiny and the assurances did not prod the scientific community to live up to Handler's characterization, Dingell instigated the creation of the Office of Scientific Integrity [later, the Office of Research Integrity (ORI)] by NIH director James Wyngaarden.

Wyngaarden proposed the office because informal

self-regulation was demonstrably inadequate for protecting the public interest in the expenditure of research funds as well as for protecting the integrity of the scientific record and the reputation of research careers.

Both offices had the authority to oversee the conduct of misconduct investigations at grantee institutions and, when necessary, to conduct investigations themselves. ORI has recently been relieved of its authority to conduct original investigations, but it can still assist grantee institutions. ORI is an effort to monitor the delegation of research and provide for the institutional conduct of investigations of misconduct allegations that, under the social contract for science, had been handled in an informal way, if at all.

The effort to ensure the productivity of research has striking parallels. In the late 1970s, Congress understood that declining U.S. economic performance might be linked to an inability of the scientific community to contribute to commercial innovation. The congressional inquiry demonstrated that different kinds of organizations, mechanisms, and incentives were necessary for the research conducted in universities and federal laboratories to have its expected impact on innovation. A bipartisan effort led to a series of laws--the Stevenson-Wydler Technology Innovation Act of 1980, the Bayh-Dole Patent and Trademark Amendments Act of 1980, and the Federal Technology Transfer Act of 1986--that created new opportunities for the transfer of knowledge and technology from research laboratories to commercial interests.

Critical to these laws was the reallocation of intellectual property rights from the government to sponsored institutions and researchers whose work could have commercial impact. At national laboratories, what the legislation called Offices of Research and

Technology Applications, which became the Office of Technology Transfer (OTT) at NIH, assisted researchers in securing intellectual property rights in their research-based inventions and in marketing them. Similar offices appeared on university campuses, contributing in some cases tens of millions of dollars in royalties to university budgets and many thousands of dollars to researchers. These changes not only allowed researchers greater access to technical resources in a private sector highly structured by intellectual property, but they also offered exactly the incentives that principal-agent theory suggests but that the social contract for science eschewed.

Such institutions as ORI and OTT spelled the end of the social contract for science, because they replace the low-cost ideologies of self-regulation and the linear model with the monitoring and incentives that principal-agent theory prescribes. Additionally, they are examples of what I call "boundary organizations" — institutions that sit astride the boundary between politics and science and involve the participation of nonscientists as well as scientists in the creation of mutually beneficial outputs. This process is "collaborative assurance."

ORI has monitored the status of allegations and conducted investigations when necessary. This policing function reassures the political principal that researchers are behaving ethically and protects researchers from direct political meddling in their work. ORI also assists grantee institutions and studies the fate of whistleblowers and those who have been falsely accused of misconduct, tapping the skills of lawyers and educators as well as scientists in this effort.

OTT has likewise employed lawyers and marketing and licensing experts, in addition to scientists, in its creation of intellectual property rights for researchers. Consequently, intellectual property has emerged as indicative of the productivity of research. Evaluators of research use patents, licenses, and royalty income to judge the contribution of public investments in research to economic goals, even as researchers use them to supplement their laboratory resources, their research connections, and their personal income.

The "collaborative assurance" at ORI and OTT demarcates a new science policy that accepts not only the macroeconomic role of government in research funding but also its microeconomic role in monitoring and providing specific incentives for the conduct of research--to the mutual purposes of ensuring integrity and productivity. Collaborative assurance recognizes that the inherited truths of the social contract for science were incomplete: A social contract for scientists is an insufficient guarantor of integrity, and governmental institutions need to supplement scientific institutions to maintain confidence in science. The public good of research is not a free good, and government/science partnership can create the economic incentives and technical preconditions for innovation.

The task for the new science policy is therefore not to reconstruct a social contract for science that was based on the demonstrably flawed ideas of a self-regulatory science and the linear model. Monitoring and incentives have replaced the trust that grounded the social contract for science. Rededication, re-articulation, and renaming do not speak to how the integrity and productivity of research are publicly demonstrated, rather than taken for granted. The new science policy should instead focus on ways to encourage "collaborative assurance" through other boundary organizations that expand the still-narrow concepts of integrity and productivity.

Ensuring the integrity of science is more than managing allegations of misconduct. It also involves the confidence of public decision-makers that the science used to inform policy is free from ideological taint and yet still relevant to decisions. Concerns about integrity undergird political challenges to scientific early warnings of climate change; the role of science in environmental, health, and consumer regulation; the use of scientific expertise in court decisions; and the openness of publicly funded research data.

The productivity of science is more than the generation of intellectual property. It also involves orchestrating research funding that targets public missions and addresses specific international, national, and local concerns, while still conducting virtuoso science. It further involves developing processes for translating research into a variety of innovations that are not evaluated simply by the market but by their contribution to other social goals that may not bear a price.

The collaborative effort of policymakers and scientists can, for example, build better analyses of environmental risks that are relevant for on-the-ground decision-makers. The experience of the Health Effects Institute, which produces politically viable and technically acceptable clean air research under a collaboration between the federal government and the automobile industry, demonstrates this concept. Not only could such boundary organizations help set priorities and conduct jointly sponsored research, but they could evaluate and retain other relevant data to help ensure the integrity of regulatory science.

Collaboration between researchers and users can mold research priorities in ways that are liable to assist both. Two increasingly widespread, bottom-up mechanisms for such collaboration are community-based research projects, or "science shops," which allow local users to influence the choice of research problems, participate in data collection, and accept and integrate research findings; and consensus conferences and citizens' panels, which allow local users to influence technological choice.

Top-down mechanisms can foster collaborative assurance as well. Expanding public participation in peer review, partially implemented by NIH, deserves broader application, particularly in other mission agencies but perhaps also in NSF. Extension services, a holdover in agricultural research from the era before the social contract for science, can serve as a model of connectivity for health and environmental sciences. The International Research Institute for Climate Prediction Research, funded by the National Oceanic and Atmospheric Administration, connects the producers of climate information with farmers, fishermen, and other end users to help climate models become more relevant and to assist in their application. Researcher-user collaborations in the extension mode can also tailor mechanisms and pathways for successful innovation even in areas of research for which market institutions such as intellectual property are lacking.

The social contract for science, with its presumption of the automatic provision of integrity and productivity, speaks to neither these problems nor these kinds of solutions. Boundary organizations and "collaborative assurance" take the first steps toward a new science policy that does.

6

IN DEFENSE OF THE SOCIAL CONTRACT

Robert Frodeman and Carl Mitcham

In January of 1803, six months before Napoleon offered him the Louisiana Territory, President Thomas Jefferson asked Congress for an appropriation of $2,500 to conduct a scientific and geographic survey of the North American West. In his letter to Congress, the president emphasized the commercial advantages of the venture: the possible discovery of a Northwest Passage and the capturing of the British fur trade. In contrast, in his personal instructions to Meriwether Lewis, Jefferson highlighted the scientific bounty of the trip: contact with unknown Indian cultures, discovery of biological and botanical wonders, and the identification of the geographic and geologic features of the region. The result of these dual charges was one of the classic journeys of exploration in U.S. history.

Almost a century and a half later, in August of 1939, scientists Leo Szilard, Albert Einstein, and others proposed a program of atomic research to President Franklin Delano Roosevelt. Although the most immediate stimulus was to beat Nazi Germany to the atomic bomb, for Szilard especially atomic physics was a means to H. G. Wells' vision of endless energy in a "world set free" from toil. Indeed, Szilard thought it might even be a way to overcome international violence by uniting all nations in a common cause greater than politics: the conquest of space.

Did the Lewis and Clark expedition embody a social contract between science and society? Or was it more the expression of a vision of the common good, with political, economic, and scientific components? Similarly, in the case of atomic energy, didn't both private industry and the government support this research in order to win the war and to advance knowledge and human welfare in a broad and mutually reinforcing synthesis?

During the past quarter-century, many have examined the relation between science and society in terms of a "social contract." Scientists, public policy analysts, and politicians have adopted such language in a largely unchallenged belief that it provides the proper framework for considering issues of scientific responsibility and the public funding of research. But neither of the cases mentioned above, nor a multitude of others that might have been chosen, involves a relationship that can be adequately described in terms of a contract. In fact, the language of a contract demeans all the parties concerned and belittles human aspirations (not to mention political discourse). Neither scientists nor citizens live by contract alone.

The social contract language is a legitimate attempt to step beyond the otherwise polarizing rhetoric of scientists and citizens in opposition. The idea of a social contract is a clear improvement over formulations that stress either the pure autonomy of science or its strict economic subservience. We believe, however, that the range of public discourse must be widened beyond that of contractual negotiation, even at the expense of opening up questions that lack simple answers. Surely human ideals demand as much attention as military security, physical health, and economic wealth, especially in a world where material achievements are greater than ever before in history. Scientists and citizens alike should strive to identify the common or complementary elements of a vision of the good, rather than discussing the quid pro quos of some illusory contract.

Ironically, social contract theory has been adopted to explain the relations between science and society just when that theory has been largely rejected as a framework for understanding politics in general. Indeed, a brief review of the rise and fall of the political philosophy of the social contract may help us appreciate the advantages and disadvantages of the notion of a social contract for science.

Social contract theory was first given modern formulation by political philosophers such as Thomas Hobbes, John Locke, and Jean-Jacques Rousseau. There are subtle distinctions among their different versions of the theory that need not concern us here. According to all versions, society originates when isolated and independent individuals make a compact among themselves to limit their freedoms in order to increase security. Before entering into their compact, individuals exist in a state of "perfect freedom" (Locke) unconstrained by any obligations to each other. The competition that results from this state of perfect freedom readily gives rise to a "war of all against all"

(Hobbes). It therefore becomes desirable to subordinate individualism for the unity of a "general will" (Rousseau).

So conceived, the contractual relationship for both politics and science presumes independent parties with divergent goals. Neither future citizens nor scientists are thought to have any ties to one another before the creation of the political or scientific compacts. Nor does either group have responsibilities to the common good. Further, by definition there are no obligations that exceed the terms of the contract, for either society or science.

These factors identify the strengths of contractual relations: They clearly protect personal freedoms and limit governmental powers. But such strengths also expose the limitations of social contract language for understanding the place of citizens and scientists in our complex and interdependent society. Surely scientists have obligations both to each other and to nonscientists prior to any formulation of mere contractual relations.

The upshot of the social contract theory was to advance the argument for human rights and the justification of enlarged democratic participation in government. The application of social contract theory to a discussion of science policy has had the similar effect of defining and protecting the rights of scientists and inviting democratic participation in the setting of broad scientific research agendas. It is nevertheless significant that such language has been largely rejected in political philosophy, for at least two interrelated reasons.

First, there is no evidence that anything like a social contract ever took place in the formation of any society. The same may be said with regard to a social contract for science. Historically, the relations that we describe ex post facto in social contract terms were never created by explicit contractual means.

Second, the social contract theory presupposes atomistic individualism as its theory of human nature, a conception that is highly problematic psychologically and sociologically. As Aristotle argued, we are fundamentally political animals in that most of the features that make us distinctly human are products of the community. Language and culture are fundamentally social rather than individual creations, although individuals obviously contribute to the furtherance of both. At the very least, the emergence of individuals is a dialectic process, involving the creation of the individual through the blending of individual initiative and communal mores.

A truer account of the science-society relationship is found in the conception of the scientific and political pursuit of the common good. Consider the issue of professional ethics in science. A scientist's ethical responsibilities are typically seen as beginning with a well-established set of obligations internal to the scientific community. Most conspicuously, these include maintaining the integrity of the research process through the honest reporting of data, fair and impartial peer review, and acknowledgement of contributions by others. But scientists also have what might be termed external obligations to avoid harming human subjects and to use their knowledge for the good.

To illustrate, compare the relation between scientists and their fellow citizens with that between physicians and their patients. When a physician saves a life, no cash payment can offer adequate compensation. One balks at describing such a relation as contractual: The nature of the exchange defies the possibility of clear and unequivocal recompense. Patients owe their physicians more than money, a fact symbolized by the social respect accorded the physician's role. Moreover, the set of obligations is reciprocal: Whereas patients and societies honor physicians, physicians take on lifetime commitments to their communities. If an illness suddenly worsens on Christmas morning, the physician must leave hearth and home. The life of the physician is closer to one of covenant and commitment than contract.

In recognition of this fact, physicians are referred to as "professionals"--that is, ones who profess or proclaim their commitment to live in accord with ideals beyond those of self-interest and the cash nexus, at least insofar as they practice medicine. One does not, for instance, expect everyone to keep confidentiality as strictly as we expect physicians to do. Similar notions of professionalism hold for lawyers, members of the military, the clergy, and engineers. The common denominator of all these practical professions is that they involve activities that go to the heart of the human condition, confronting matters that lie beyond the prosaic: issues of life and death, freedom, justice, and security.

Curiously, however, scientists are seldom denominated professionals in quite the same way. Scientists may be thought of as theoretical professionals. Scientific societies have, for instance, been slower than medical, legal, or engineering societies to adopt professional codes of ethics that increasingly affirm social responsibility above and beyond any contractual determinations. Moreover, when scientists are called professionals, this is often done to promote an independence that may be at odds with the social good, thus calling for qualification.

The good in science, just as in medicine, is integral to and finds its proper place in that overarching common good about which both scientists and citizens deliberate. Politics in this sense is more than the give and take of interest groups. Instead, it is that reflective process by which citizens make informed choices on matters concerning the shared aspects of their lives. Politics denotes the search for a common good, where people function as citizens rather than only as consumers.

From this perspective, the good intrinsic to science consists not only in procedures that are designed to preserve scientific integrity. It also expands into the goods of knowledge, of the well-ordered life, of fellowship and community, and the wonder accompanying our understanding of the deep structure of things. Indeed, there are even aesthetic and metaphysical dimensions of the good in scientific research. During the moon landings, the beauty of Earthrise over the lunar landscape and the collective sense of transcendence we felt in watching humans step out onto another world may well have been the enduring legacy of the moon missions, rather than any of the varied economic and technological spinoffs.

Although few politicians would admit to voting for a scientific project on its aesthetic or metaphysical merits alone, much of science has precisely such results. Contract economics must not be allowed to crowd out recognition of more expansive but absolutely fundamental motivations. Reductionism may not be a sin in science, but it is in politics.

Conceived under the sign of the common good, scientists have much broader obligations than those of simple scientific integrity. Indeed, even internal obligations find more generous and inclusive foundations in the notion of the common good than in the language of social contract. From the perspective of the common good, it is incumbent upon the scientist to preserve the integrity of science, treat all experimental subjects with respect, inform the community about research under consideration, provide ways for the community to help define the goals of scientific research, and report in a timely manner the results of the research in forums accessible to the non-specialist.

The shift from thinking of science as involved in a social contract to science as one aspect of a continuing societal debate on the common good broadens science policy discourse. It also deepens reflection on the science-society relation in science and in politics.

One major limitation of the idea of a social contract for science is that it has implications only for publicly funded science. Science policy discussions emphasizing social contract language exclude a large segment of the scientific community not funded by government. Questions focusing on the common good (without denying important distinctions between privately and publicly funded science) will include concerns of a far larger constituency. For instance, shouldn't we be asking questions about the goodness of human cloning, not simply whether the tax dollars of those who oppose human cloning should be used to fund it?

For scientists themselves, working in both the private and public sectors, trying to articulate a common good will point beyond justifications of science merely in terms of economic benefit. What science can bring to society are not just contractual benefits but enhanced intelligence and even beauty. Using the language of the common good, scientists will be encouraged to make a case for science as a true contributor to culture. E. O. Wilson's defense of biodiversity through "biophilia" and his notion of a "consilience" between science and the humanities are but one salient expression of such an approach.

Using the framework of the common good opens science to being delimited by other dimensions of human experience. Science is not the whole of the common good, and as part of that whole it may sometimes find its work restricted in order to serve more inclusive conceptions of the good life. Liberal democratic societies restrict experimentation on human subjects because of a good that overrides any scientific knowledge that may result from such work. But surely this is a vulnerability that science can survive, and should affirm. This approach will also help take public discussion out of the framework of a hackneyed contest between reason and revelation, as in the evolution versus creationism controversy. It is possible, after all, to have a reasonable discussion on the nature of the good life without constant reference to the facts of science or the claims of fundamentalism.

Given the extraordinary effects of scientific discoveries and technological inventions during the 20th century, effects that will only increase throughout the 21st century, social contract theory cannot give a sufficiently comprehensive account of the science-society relationship. Scientists, like all their fellow citizens, must be concerned not just about advancing their own special interests but more fundamentally about the common good. This broader obligation is operative on two levels: that of internal professional responsibility and that of citizenship.

Professional responsibilities commonly described as internal as well as external have become unavoidable for the scientist today, especially for the scientist employed or supported by federal money. Appreciating such demands—acknowledging the claims of community without compromising the integrity of the scientific process—has become a central issue for practicing scientists and for those engaged in science education.

This "trans-contractual" view of scientific responsibility challenges the long-held belief that the integrity of the scientific process is founded on the exclusive allegiance to facts and the banishment of values of any kind. It has been an honored principle that scientists qua scientists must not attempt to draw political conclusions from their scientific research and that their work should be isolated from political pressures of all types. With scientists recognizing their own citizenship, and citizens realizing the scientific fabric of their lives, these claims become tenuous.

First, as the social contract language recognizes, there are inevitable, and increasing, places where the scientist and the public interact. This is a result of a wide set of changes in society. These changes include the loss of a clear consensus about societal goals with the end of the Cold War and the state of emergency that it fostered, more rigorous standards of accounting for the spending of public monies, and the increasing relevance of scientific data to various types of environmental questions and controversies. This means that scientists' responsibilities include understanding the concerns of the public as well as being able to explain their work to the community.

Second, truism of recent philosophy of science is that, although the scientific research process can and must be fair, the full exclusion of values is an unattainable and even undesirable goal. Human interests are always tied to the production of knowledge. The collection and interpretation of data are constrained by a variety of factors, including limitations of time, money, and expertise. The most rigorous objective scientific procedure is motivated by personal or social values, whether they be economic (generating profits or gaining tenure), political (nuclear deterrence or improving community health), or metaphysical (the love of understanding the deep nature of things). Finally, various types of methodological value judgments inevitably come into play, such as the perspectives one brings on the basis of past experience and training.

The social contract language has arisen in an attempt to take account of such factors. But it is only the principle of the common good that can do full justice to them. In 1970, during testimony before Congress, A. Hunter Dupree, an esteemed historian of U.S. science and government policy, called for the creation of a new kind of Manhattan Project. The World War II Manhattan Project had brought together a spectrum of atomic scientists and engineers to create the atomic bomb. Dupree's new Manhattan Project "would do away with the conventional divisions between the natural and social science and humanities, and by drawing on people from many disciplines . . . would provide the enrichment and stimulation of unaccustomed patterns." The goal of such a pluralistic project might well be described as a full and rich articulation of the common good, for scientist and nonscientist alike.

Dupree opened his testimony with a quotation from John Wesley Powell, one of the founders of public science in the United States and the second director of the U.S. Geological Survey. Charles Groat, who was director of the USGS from 1998-2005, once echoed Dupree by calling for a future in which science "is more cooperative, more integrated, and more interdisciplinary." It is not the refining or renegotiation of a contract that will lead in this direction, but public discussion of the common good of science and of society.

7

TECHNOLOGY AND SOCIAL TRANSFORMATION

Michael Crow and Daniel Sarewitz

Technological innovation sustains a fundamental tension of civilization, the tension between humanity's quest for more control over nature and the future, and our equally strong desire for stability and predictability in the present. The original Luddites were not against technology per se. They were against losing their jobs, and so they smashed the power looms that had put them out of work. The change wrought by technological advance continually remakes society, and this transformational process is on the one hand central to the dynamic that is commonly labeled "progress," and yet on the other is a source of continual destabilization and dislocation as experienced by individuals, communities, institutions, nations, and cultures.

In the age of science and technology (S&T), the federal government has increasingly become the prime catalyst for scientific advance and technological innovation. At the same time, modern government is also continually responding to and managing the transformational power of science and technology. Yet there is little effort to understand the relation between these two critical activities—advancing knowledge and innovation, and responding to their impact or to linking them in ways that can enhance the value and capability of each.

A single technological innovation can remake the world. When the metal stirrup finally migrated from Asia to western Europe in the 8th-century, society was transformed to its very roots. For the first time, the energy of a galloping horse could be directly transmitted to the weapon held by the man in the saddle—a combat innovation of devastating impact. Because horses and tack were costly, they were possessed almost exclusively by landowners. Battlefield prowess and wealth were thus combined, and from this combination grew not just the traditions of a "warrior aristocracy" but the structure of European feudal society itself. Later, when the Anglo Saxon King Harold prepared to defend Britain against the invading Normans in 1066, he actually dispensed with his horse and ornamental wooden stirrups, choosing to lead his numerically superior forces on foot. The outnumbered Normans, as the historian Lynn White documented in his classic, *Medieval Technology and Social Change*, boasted a strong, stirrup-equipped cavalry, and thus won the day—and the millennium.

Such a narrative has the ring of mythology, yet the experience of the industrialized world reinforces the knowledge that a new machine can help change everything. The invention of the cotton gin in the late 18th century allowed a vast expansion of cotton cultivation in the American south—and directly fueled a resurgence in the importation and use of slaves for plantation labor. One hundred and fifty years later, the mechanical cotton picker suddenly rendered obsolete the jobs of millions of African American sharecroppers, and catalyzed a 30-year migration of five million people out of the rural south and into the cities of the north. While the development of the mechanical cotton picker was no doubt inevitable, its proliferation was consciously accelerated by plantation owners who, fearing the rise of the civil rights movement, sought quickly to find a technological replacement for the existing system of exploitation of labor upon which they were economically dependent.

These examples point not only to the power of new technologies to transform society, but to the comprehensive interconnectedness of technological change and the complex social structure of society. The invention of the stirrup as a battlefield tool was in some very intricate way connected to the development and expansion of feudalism in Europe; the evolution of agricultural technology for a single cash crop is indissolubly bound to the ongoing struggle to overcome the U.S. legacy of slavery, segregation, and bigotry. More familiarly, a single class of technology—nuclear weapons—was a central determinant of geopolitical evolution after the end of World War II. Cars, television, air conditioning, and vaccinations have all stimulated foundational changes in society during the past century.

New technologies rarely emerge in isolation. The industrial revolution is not just the story of harnessing steam power to factory production capability, but also the story of technological revolutions in transport, communication, construction, agriculture, resource extraction, and, of course, weapons development. These technological systems penetrated the innermost niches of society—home and family, school, workplace, community—and forced them to change. They also introduced completely new social phenomena, and stimulated the invention of completely new institutions.

The industrial revolution created the macroeconomic phenomenon of unemployment. Prior to the 19th century, even the most economically and politically advanced societies were dominantly agrarian and rural. For the majority of people, work was rooted in the home and the family. Vagaries of weather and transportation imposed irregularities and hardship, but most people and families harbored a diversity of skills that gave them independence from the marketplace and resilience to cope with a variety of challenges. In hard times, resort to subsistence farming and barter was usually possible.

Industrialization and urbanization linked workers far more closely to the larger economic market, while removing the need and ability for them to maintain the diverse skills necessary for survival in the pre-industrial world. The traditional connection between manufacturing and agriculture in the home was sundered by new economic organization and by geography. Labor itself became a commodity, subject to the same fluctuations and influences as other commodities. During an economic downturn, factories fired people or closed down entirely. For the first time, workers could not easily respond to changing economic conditions by switching to a different type of work or moving to a subsistence mode. Karl Polanyi observed in *The Great Transformation: The Political and Economic Origins of Our Time*: "To separate labor from other activities of life and to subject it to the laws of the market was to annihilate all organic forms of existence and to replace them by a different type of organization, an atomistic and individualistic one."

As technological innovation interacts with society to create new phenomena, such as unemployment, society also responds by developing new types of institutions and response mechanisms. Today we can recognize the problem of unemployment as central to a diversity of social, political, and economic structures and activities ranging from the organization of labor to insurance safety nets to educational programs to immigration policy. Unemployment rates are a key indicator of economic health, and a key determinant of political behavior. National and international economic policies focus strongly on managing unemployment, even as theoretical investigations seek to clarify the relation between unemployment rates and other key attributes of modern economies.

The general point is that transformational technology represents one variable in a complex assemblage of dynamic, interrelated societal activities. Decision making processes tend to address each of these activities in isolation from the others, e.g., conduct of research and development (R&D), dissemination of innovation products, development of regulations, reform of institutions. Concerted action occurs when a given innovation stimulates enough transformation to demand a response from other sectors of society. This response then triggers additional changes, which in turn demand further modulation. The process is reactive, discontinuous, disruptive, and sequential — like billiards. The challenge is to move toward a process of technology supported societal progress where different sectors and activities can continually coevolve in response to knowledge about one another's needs and constraints — like an ecosystem. We are not there yet.

A brief consideration of evolution of information technologies helps to bring this look at societal transformation into the present. Gutenberg's perfection of the printing press of course had enormous transformational impact, allowing the broad dissemination of written texts and consequent expansion of information — and literacy — that undermined the Church's hegemony over knowledge and culture, and helped promote the dissolution of medieval social structure. Lewis Mumford suggested that the printed word represents "the media of reflective thought and deliberate action," a prerequisite, perhaps, for the intellectual achievements of the Enlightenment. But he also observed as early as 1934 that new modes of electronic communication were increasing the speed of information exchange to levels that made reflection impossible, and increasing the volume of information transmission to a point that exceeded our absorptive capacity.

The implications on democracy itself are far from clear. On the one hand, proliferation of information dissemination networks means greater access by more people to more information — and a greater capacity to communicate one's ideas and preferences in democratic fora.

Control of information by authoritarian governments is becoming increasingly futile, and organization of democratic opposition increasingly enhanced, by new information technologies. But when this same capacity translates into 10,000 identical e-mail messages sent to a Member of Congress in support of a particular bill, one is hard-pressed to suggest that democracy is the beneficiary.

Of particular concern is the recent increase in public referenda aimed at bypassing the legislative process. The barriers to putting referenda on ballots have been enormously reduced by information and communication technologies that can be used to disseminate ideas and organize group action with relatively little effort. While on the one hand this type of direct democracy can be a refreshing antidote to sclerotic legislative process, on the other it is quite often devoid of any serious deliberative process or public discourse, reflecting perhaps the pique of one well-organized interest group or individual, and the substantiation of a Warholian politics where anyone with access to a decent list-serve can lead a movement for a day. Is democracy in transition?

The implications of the revolution in information and communications on the distribution of economic benefits in society are also problematic. Does the troubling increase in wealth concentration that characterizes both the U.S. and the global economy derive from the way that advanced technologies diffuse in market economies? Does the synergistic character of information and communication networks mean that disenfranchised populations and nations will find it increasingly difficult to participate in the spectacular economic growth that we have seen in the past decade? In other words, are the benefits of technology becoming increasingly appropriable by particular sectors of society, and is this in part an attribute embodied in new types of technological systems? Society is ill-prepared to answer such questions, let alone act on them in a knowledgeable manner.

Paradoxically, these concerns cut both ways. In the information society, the increasing ease of information dissemination may also threaten our system for protecting intellectual property and innovation. From pirated CD's sold on the streets of Shanghai to the advent of Napster, the concept of intellectual property seems increasingly vulnerable. Are we looking to a future where such protection is no longer practically possible? Does a world without patents and copyrights seem unimaginable? More unimaginable than, say, the loss of monopoly over the written word would have seemed to the Church in 1450?

At issue here is not the value of change, but the path that change follows. What may look in retrospect like the march of progress may be experienced in real time as wrenching dislocation. The Dickensian squalor of 19th century London remains a symbol of the human impacts of technological change. Faced with unprecedented societal transformations, the English government (as well as other European states) failed to develop effective policies that could accommodate the rapid transition from rural agrarian to urban industrial society. Today, the plight of many overpopulated developing nations is the post-industrial, global manifestation of the same failure.

We see the fingerprints of societally-transforming technological systems in the controversy over genetically modified organisms; in the 40 million Americans with no medical insurance; in the general inability of our public school systems to create a citizenry able to take advantage of the opportunities of the knowledge economy; in the challenges presented by the aging of our population; in the rising atmospheric carbon dioxide levels that reflect 150 years of industrial dynamism.

Even the unprecedented rise of civil and ethnic conflict throughout the world in the past decade can be plausibly connected to technological transformation. Approaching this phenomenon from entirely different directions, the political scientists Samuel Huntington and Benjamin Barber each conclude that advanced communication and information technologies have created new fora for expressing ethnic identity and pursuing and strengthening cultural solidarity. Virtual communities, for example, can act to maintain identity over great distance, while also more efficiently garnering resources to support the expression of cultural goals. As Barber observes: "Christian Fundamentalists [can] access Religion Forum on CompuServe Information Service while Muslims can surf the Internet until they find Mas'ood Cajee's Cybermuslim document." The result may be locally empowering and globally divisive.

The marriage of science and technology beginning in the latter part of the 19th century accelerated the process of innovation, and thus the process of societal transformation as well. If the industrial revolution played itself out in less than 200 years, the electronics revolution seems likely to have a working life of at least 75 years. The biotechnology revolution, while hardly on its feet, is already expected to morph into the nanotechnology revolution in the next 50 years. What type of transformations might this revolution bring about?

Our point here is not to predict the future of nanotechnology and its impacts — an impossible task — but to illustrate the direction and scale of thinking that will be necessary if we are to successfully manage the interaction of new knowledge and innovation with society. Judging by the literature prepared by the government, as well as the work of futurists and other techno-pundits, the promise of nanotechnology to re-make our world seems virtually infinite.

So if nano-technology is going to revolutionize manufacturing, health care, travel, energy supply, food supply, and warfare, then it also will transform labor and the workplace, the medical system, the transportation and power infrastructure, the agricultural enterprise, and the military. Each one of these technology-dependent sectors is operated by and for human beings, who act within institutions and cultures, according to particular regulations, norms, and heuristics, all of which may reflect decades or even centuries of evolution, negotiation, and tradition. Not one of them will be "revolutionized" without significant difficulty. The current chaos in our health care system is emblematic of this type of difficulty.

In the near term, the current state of knowledge may suggest that the first wave of useful nano-technologies will lie in the area of detection and sensing. The capacity to detect, precisely identify, and perhaps isolate single molecules, viruses, or other complex, nano-scale structures has broad application in such areas as medical diagnosis, forensics, national defense, and environmental monitoring and control. The potential for direct benefits is obvious; how might this evolving capacity influence society?

When detection outpaces response capability — as it often does — ethical and policy dilemmas inevitably arise. For example, it is already possible to identify genetic predisposition to certain diseases for which there are no known cures, or to diagnose congenital defects in fetuses for which the only cure is abortion. In the environmental realm, new technologies that detect pollutants at extremely low concentrations raise complex questions about risk thresholds and appropriate remediation standards. The presence of tiny amounts of toxic materials in ground- water may justifiably raise alarm among the public even if the health risk cannot be assessed, and the technological capacity for remediation does not exist. These types of dilemmas may be expected to accelerate and proliferate with the advance of nano-detection technologies.

Innovations in sensing and detection may transform existing societal mechanisms and institutions that were designed to cope with uncertainty and incomplete or imprecise information. The insurance industry, for example, deals with incomplete knowledge about the health of specific individuals by spreading its risk among large populations. If there is no way to distinguish between someone who is going to suffer a potentially lethal middle-age heart attack, and someone who is going to live to 105, then they can both get health and life insurance. Society clearly gains from this arrangement: costs are broadly disseminated, and benefits are delivered to those who most need them.

Medical sensors that can, for example, detect an array of medically relevant signals at high sensitivity and selectivity promise to aid diagnosis and treatment of disease, but also to develop predictive health profiles of individuals. Today, health and life insurance companies often use pre-existing conditions as a basis for denying or restricting coverage. The advent of nano-detection capabilities will considerably expand the information that insurance companies will want to use in making decisions about coverage. The generation of new information might thus destabilize the risk-spreading approach that allows equitable delivery of social benefits to broad populations. How will society respond?

Nano-technology offers a dizzying range of potential benefits for military application. Recent history suggests that some of the earliest applications of nanotechnology will come in the military realm, where specific needs are well-articulated, and a customer—the Department of Defense—already exists. One area of desired nano-innovation lies in the increased use of enhanced automation and robotics to offset reductions in military manpower, reduce risks to troops, and improve vehicle performance. How might progress in this realm interact with the current trend toward rising civilian casualties (in absolute terms and relative to military personnel) in armed conflict worldwide? As increased robotic capability is realized in warfare, will we enter an era when it is safer to be a soldier in wartime than a civilian?

Such considerations are simple extrapolations of current trends in technological innovation and societal transformation. More adventurous speculation is tempting but is perhaps best confined to science fiction novels. The question of public response to nano-innovation, however, should not be avoided, even at this early stage. The ongoing experience of public opposition to old technologies such as nuclear power, new technologies such as genetically modified foods, and prospective technologies such as stem cell therapies, needs to be viewed as integral to the relationship between innovation and societal transformation.

Three observations are particularly relevant here. First, the impact of rapid technological innovation on people's lives is usually not consensual. Second, in the short term at least, the social changes induced by new technologies usually create both winners and losers (where what is lost may range from a job to an entire community). Third, rapid technological change can threaten the social structure, economic stability, and spiritual meaning that people strive in their lives to achieve. As the nanotechnology revolution begins to unfold in all its promise and diversity, such issues are bound to express themselves. They should not be viewed as threats, or as manifestations of intellectual weakness or repugnant ideology. Rather, they need to be recognized as a central part of the human context for technological change.

When resources are allocated for R&D programs, the implications for complex societal transformation are not considered. The fundamental assumption underlying the allocation process is that all societal outcomes will be positive, and that technological cause will lead directly to a desired societal effect. The literature promoting the National Nanotechnology Initiative expresses this view. The current policy approach thus addresses two elements: the conduct of science and technology and the products of science and technology.

These elements reflect the internal workings of the R&D enterprise. The fact that societal outcomes are not a serious part of the framework seems to derive from two beliefs: (1) that the science and technology enterprise has to be granted autonomy to choose its own direction of advance and innovation; and (2) that because we cannot predict the future of science or technological innovation, we cannot prepare for it in advance. These are oft-articulated arguments, not straw men. Yet the first is contradicted by reality, and the second is irrelevant. The direction of science and technology is in fact dictated by an enormous number of constraints (only one of which is the nature of nature itself). And preparation for the future obviously does not require accurate prediction; rather, it requires a foundation of knowledge upon which to base action, a capacity to learn from experience, close attention to what is going on in the present, and healthy and resilient institutions that can effectively respond or adapt to change in a timely manner.

If we stand the current S&T policy approach on its head, and start by thinking about desired social outcomes, rather than more money for the R&D enterprise, where would we begin? We might identify several very general categories of outcomes that most people would agree are worth thinking about. For example:

- Social equity: the distribution of the benefits of science and technology.

- Social purpose: the actual goals of societal development that we want to pursue or advance.

- Economic and social enterprises: the shape and make-up of the institutions at the interface between technology and the human experience.

How can consideration of these types of outcomes be integrated into the S&T policy framework? The years since World War II have seen a very gradual evolution in the effort to connect thinking about S&T to thinking about the outcomes of S&T in society. A science policy report issued by the Truman Administration, for example, mentioned in its first pages the need to prepare for both the positive and negative impacts of scientific and technological change. The rise of the environmental movement in the late 1960s reflected a public demand that society devote more S&T resources to the achievement of desired social outcomes like clean air and water. The creation of the U.S. Office of Technology Assessment reflected growing public concern about the need to understand the societal implications of technological choices. Over the past decade, federally funded programs on the human dimensions of global climate change, and the ethical, legal, and social implications of the human genome project and information technologies, have been supported as adjuncts to much, much larger core research agendas in the "hard" sciences.

Yet S&T policy itself remains input-driven. Concepts such as sustainability, and analytical tools such as human development indicators, provide conceptual frameworks for linking R&D to societal outcomes, and in fact imply that outcomes are to some degree implicit in the choices we make about R&D inputs. These types of insights point the way toward the next step: to implement an approach to R&D policy that addresses the complex interconnections between technological advance and societal response. Such an approach would need to integrate the pursuit of innovation with an evolving understanding of how innovation and society interact, and include mechanisms to feed this understanding back into the innovation process itself. (In a very specific way, the private sector does this as a matter of course, as it uses consumer input to continually refine and improve the next generation of products.)

To be serious about preparing for the transformational power of a coming nanotechnology revolution, policymakers would need first to get serious about developing knowledge and tools for more effectively connecting R&D inputs with desired societal outcomes. This in turn would require the creation of a dedicated intellectual, analytical, and institutional capability focused on understanding the dynamics of the science-society interface and feeding back into the evolving nanotechnology enterprise. Such a capability should include the following elements:

Analysis of past and current societal responses to transforming technologies

A case history approach could be used to investigate the diverse avenues that society has followed in responding to a range of technological advances. Understanding the roles and relations between the media, academia, policy makers, institutions, and cultural factors could be the basis for assessing — and anticipating — the likely trajectories of technology-induced social change.

Comprehensive, real time assessment and monitoring of the nano-science and nano-technology enterprise

At this relatively early stage, it should be feasible to build a database of important activities in nanotechnology, and then track the evolution of the enterprise over time, in terms of directions of research and innovation, resources used, public and private sector roles, publications and patents, marketed products, and other useful indicators. This type of information is essential to understanding potential impacts.

A science communication initiative, to foster dialogue among scientists, technologists, policy makers, the media, and the public

Understanding, tracking, and enhancing the processes by which information about nanotechnology diffuses from the laboratory to the outside world is central to understanding the social transformation process as it occurs. Of equal importance is the need to understand and monitor how public attitudes and needs evolve, and how they reach back into the innovation system. Empirically grounded, research-based inquiries on communication can be the basis for strategies to improve social choice in ways likely to secure favorable outcomes.

A constructive technology assessment process, with participants drawn from representatives of the R&D effort, the policy world, and the public

Technology assessment is both a process for bringing together a range of relevant actors, and an evolving product that can inform and link the innovation and decision-making processes. Understanding the changing capabilities of both the nanotechnology enterprise and various sectors and institutions likely to be affected by the enterprise can contribute to a healthy policy-making environment where innovation paths and social goals are compatible and mutually reinforcing. Should nano-science and nano-technology yield even a small proportion of their anticipated advances, the impacts on society will be far-reaching and profound — as revolutionary as the introduction of electricity, piped water, antibiotics, or digital computing.

We can allow these transformations to surprise and overwhelm us, and perhaps even threaten the prospects for further progress. Or we can choose to be smart about preparing for, understanding, responding to, and even managing the coming changes, in order to enhance the benefits and reduce the disruption and dislocation that must accompany any revolution.

agencies and in Bush administration policies. A February 2004 "Scientist Statement on Restoring Scientific Integrity to Federal Policy Making," initially signed by 62 scientists, many with reputations as national leaders in their fields, summarized the charges: "When scientific knowledge has been found to be in conflict with its political goals, the administration has often manipulated the process through which science enters into its decisions ... The distortion of scientific knowledge for partisan ends must cease if the public is to be properly informed about issues central to its well-being, and the nation is to benefit fully from its heavy investment in scientific research." The issue made it into the 2004 presidential campaign, where Democratic candidate John Kerry pledged: "I will listen to the advice of our scientists, so I can make the best decisions.... This is your future, and I will let science guide us, not ideology."

The allegations were neatly summarized in the title of Chris Mooney's 2005 book, *The Republican War on Science*. In the run-up to the 2008 presidential election, a Democratic party website promised: "We will end the Bush administration's war on science, restore scientific integrity, and return to evidence-based decision-making."

In this context, restoring science to its rightful place became good politics. But an urgent question remains: how do we know the "rightful place" when we see it?

One way to search for an answer is to look at the Obama administration for indications of what it has done differently from the Bush administration in matters of science policy. The obvious first candidate for comparison would be embryonic stem cell research, which for many scientists and members of the public symbolized President Bush's willingness to sacrifice science on the altar of a particularly distasteful politics:

pandering to the religious right's belief that the sanctity of human embryos outweighs the potential of stem cell research to reduce human suffering.

When President Bush announced his stem cell policy in August 2001, he tried to walk a moral tightrope by allowing federal support for research on existing stem cell lines, thus ensuring that no embryos would be destroyed for research purposes, while loosening the ban on embryo research that Congress had established in 1994, though on a very limited basis. The president reported that there were about 60 such existing cell lines, a number that turned out, depending on one's perspective, to be either overly optimistic or a conscious deception; the actual number was closer to 20.

Lifting the restrictions on stem cell research was a part of the 2004 campaign platform of Democratic presidential candidate John Kerry, as it was of Barack Obama four years later. Less than two months into his presidency, Obama announced that he would reverse the Bush policies by allowing research on cell lines created after the Bush ban. The president instructed the director of the National Institutes of Health (NIH) to "develop guidelines for the support and conduct of responsible, scientifically worthy human stem cell research."

In announcing the change, President Obama emphasized the need to "make scientific decisions based on facts, not ideology," yet the new policy, as well as the language that the president used to explain it, underscores that the stem cell debate is in important ways not about scientific facts at all, but about the difficulty of balancing competing moral preferences. The new policy does not allow unrestricted use of embryos for research or the extraction of cell lines from embryos created by therapeutic cloning. In explaining that "[m]any thoughtful and decent people are conflicted about, or strongly oppose, this research," President Obama was acknowledging that, even in its earliest stages, the small group of cells that constitute an embryo are in some way different from a chemical reagent to be sold in a catalog or an industrially synthesized molecule to be integrated into a widget.

To protect women from economic and scientific exploitation, and in deference to the moral and political ambiguity that embryos carry with them, no nation allows the unrestricted commodification of embryos, and some, including Germany, have bans on destroying embryos for research purposes. Although most Americans favor a less restrictive approach to stem cell research than that pursued by President Bush, the issue is inherently political and inherently moral. Thus, some of the cell lines approved for research under the Bush restrictions might actually not be approved under the Obama guidelines because they may not have been obtained with the appropriate level of prior informed consent of the donor, a moral constraint on science that apparently did not concern President Bush.

Shortly after President Obama laid out his new approach, a *New York Times* editorial accused him of taking "the easy political path" by allowing federal research only on excess embryos created through in vitro fertilization. The accusation was ambiguous; it implied either that there is a "hard" political path, or that there is a path that is entirely nonpolitical. Given the state of public opinion, apparently President Bush took the hard political path and paid the political price.

But the idea that there is a path beyond politics, one that is paved with "facts, not ideology," is false — indeed, itself a political distortion — so long as significant numbers of people see human embryos as more than just a commodifiable clump of molecules. Moreover, there is nothing at all anti-science about restricting the pursuit of scientific knowledge on the basis of moral concerns. Societies do this all the time; for example, with strict rules on human subjects research. The Bush and Obama policies differ only as a matter of degree; they are fundamentally similar in that neither one cedes moral authority to science and scientists. When it comes to embryonic stem cells, the "rightful place of science" remains a place that is located, debated, and governed through democratic political processes.

Another common allegation about Bush administration abuse of science focused on decisions that ignored the expert views of scientists in deference to pure political considerations. An early signal of how President Obama will deal with apparent conflicts between expert scientists and political calculus came when the president decided to slash funding for the Yucca Mountain nuclear waste repository, the only congressionally approved candidate for long-term storage of high-level nuclear waste. Since the late 1980s, the Department of Energy has spent on the order of $13 billion to characterize the suitability of the 230-square-mile site for long-term geological storage, probably making that swath of Nevada desert the most carefully and completely studied piece of ground on the planet. At the same time, because of the need to isolate high-level waste from the environment for tens of thousands of years, uncertainty about the site can never be eliminated. Writing in *Science* magazine about the issue, two respected geologists and experts on the subject, Isaac Winograd and Eugene Roseboom, explained that this persistence of uncertainty is inherent in the nuclear waste problem itself, not just at Yucca Mountain, and that uncertainties can best be addressed through a phased approach that allows monitoring and learning over time. Winograd and Roseboom suggest that the Nevada site is suitable for such an approach, echoing a recent report of the National Academies. But they also emphasize that the persistence of uncertainties "enables critics … to ignore major attributes of the site while high-lighting the unknowns and technical disputes."

Among those critics have been the great majority of the citizens of Nevada, a state that has acted consistently and aggressively through the courts and political means to block progress on the site since it was selected in 1987. A particularly effective champion of this opposition has been Harry Reid, majority leader in the U.S. Senate and one of the most influential and powerful Democratic politicians in the nation. Now add to the mix that Nevada has been a swing state in recent presidential elections, supporting George Bush for president in 2000 and 2004, and Bill Clinton in 1992 and 1996. President Bush strongly supported the Yucca Mountain site, as did 2008 Republican candidate John McCain. All of the major Democratic presidential candidates, seeking an edge in the 2008 election, opposed the site; shutting it down was one of Barack Obama's campaign promises, which he fulfilled by cutting support for the program in the fiscal year 2010 budget, an action accompanied by no fanfare and no public announcement.

At this point it is tempting to write: "It's hard to imagine a case where politics trumped science more decisively than in the case of Yucca Mountain, where 20 years of research were traded for five electoral votes and the support of a powerful senator," which seems basically correct, but taken out of context it could be viewed as a criticism of President Obama, which it is not.

But the point I want to make is only slightly more subtle: Faced with a complex amalgam of scientific and political factors, President Obama chose short-term political gain over longer-term scientific assessment, and so decided to put an end to research aimed at characterizing the Yucca Mountain site. This decision can easily be portrayed in the same type of language that was used to attack President Bush's politicization of science. John Stuckless, a geochemist who spent more than 20 years working on Yucca Mountain, was quoted in *Science* making the familiar argument: "I think it's basically irresponsible. What it basically says is, they have no faith in the [scientists] who did the work … Decisions like that should be based on information, not on a gut feeling. The information we have is that there's basically nothing wrong with that site, and you're never going to find a better site."

To this observer, it turns out that the nostrums and tropes of the Republican war on science are not easily applied in the real world, at least not with any consistency. The "rightful place" of science, in short, is hard to find. Or perhaps we are looking for it in all the wrong places? When President Obama was urgently seeking to push his economic stimulus package through Congress in the early days of his administration, he needed the support of several Republican senators to guard against Republican filibuster and to bolster the claim that the stimulus bill was bipartisan. Senator Arlen Specter, who suffers from Hodgkin's disease, agreed to back the stimulus package on the condition that it included billions of dollars in additional funding for NIH. For this price a vote was bought and a filibuster-proof majority was achieved.

Now there is nothing at all wrong with making political deals like this; good politics is all about making deals. What's interesting in this case is the pivotal political importance of a senator's support for science. If Senator Specter (who, perhaps coincidentally, underwent a party conversion several months later) had asked for $10 billion for a new weapons system or for abstinence-only counseling programs, would his demand have been met? In promoting the stimulus package to the public, one of the key features highlighted by congressional Democrats and the Obama administration was strong support for research, including $3 billion for the National Science Foundation, $7.5 billion for the Department of Energy, $1 billion for the National Aeronautics and Space Administration, and $800 million for the National Oceanic and Atmospheric Administration, in addition to the huge boost for NIH.

These expenditures are on one level certainly an expression of belief that more public funding for research and development is a good thing, but they are also a response to the discovery by Democrats during the Bush administration that supporting science (and, equally important, accusing one's Republican opponents of abusing or undermining science) is excellent politics; that the position appeals to the media and to voters and is extremely difficult to defend against. Democrats were claiming not simply that money for science was good stimulus policy but that it was a necessary corrective to the neglect of science under the Bush administration. Speaker of the House Nancy Pelosi quipped: "For a long time, science had not been in the forefront. It was faith or science, take your pick. Now we're saying that science is the answer to our prayers."

Is money for science good stimulus policy? Experts on economics and science policy disagreed about whether ramming billions of dollars into R&D agencies in a short period of time was an effective way to stimulate economic growth and about whether those billions would be better spent on more traditional stimulus targets such as infrastructure and increased unemployment benefits. Lewis Branscomb, one of the nation's most thoughtful observers of U.S. science policy, summed up the dilemma in a University of California, San Diego, newsletter article: "If the new research money is simply spread around the academic disciplines, it will be great for higher education, but will be a long time contributing to national problem-solving." And beyond the stimulus question, were such sharp increases in science funding good science policy? Writing in *Nature*, former staff director for the House Science Committee David Goldston observed that "A stimulus bill is not the ideal vehicle for research spending, and if scientists and their proponents aren't careful, the bill is a boon that could backfire." Goldston highlighted three concerns: first, "that being included in the stimulus measure could turn science spending into a political football," second, that "a brief boom could be followed by a prolonged bust," and "third, and perhaps most troubling … that inclusion in the stimulus bill means the science money must be awarded with unusual, perhaps even reckless, speed."

As a matter of national politics, however, the immediate benefits were obvious for the president and for Democratic politicians. Democrats are finally discovering a politically powerful symbol of what they want to stand for, a symbol that captures the American reverence for progress and exemplifies a positive role for government that cannot easily be tarred by Republicans as "tax-and-spend" or anti-market but on the contrary is widely believed by Americans to be the key to a better tomorrow. Consider, for example, the complete sentence that President Obama used in his first inauguration speech: "We will restore science to its rightful place and wield technology's wonders to raise health care's quality and lower its costs."

Science in its "rightful place" is linked to the curing of disease and the reduction of health care costs. Who could be against such things? They stand in for a more general belief in human progress. Never mind that President Obama's claim—that more medical technology created by more scientific research will reduce health care costs—is, to put it mildly, implausible. After all, continual scientific and technological advance appears to increase health-care costs, not reduce them. New medical knowledge and technology will undoubtedly relieve suffering and extend life for many, and it will probably reduce costs in some cases when it replaces current types of care, but the overall effect of progress in medical science and technology will be the same as it as been for decades: to increase total health care spending. But a small inconsistency is the hobgoblin of policy wonks; surely the key point is that science, linked to progress, is change we can all believe in.

Perhaps the best way to understand what seems to be happening to science as a political symbol for Democrats is to consider, in contrast, the value of "national defense" as a political symbol for Republicans. President Bush made powerful use of the idea that Republicans are more concerned about national security, and more able to protect it, than are Democrats, both in justifying his prosecution of the war in Iraq and in attacking John Kerry during the 2004 election campaign. In the 1980 presidential campaign, Ronald Reagan made devastatingly effective use of the notion that President Carter was soft on defense, and a signal priority for the Reagan administration from its earliest days was to greatly increase expenditures on the military, just as President Obama has done for science.

Because "national security" and, it now turns out, "science" are tropes that resonate powerfully with significant parts of the voting public, they make highly potent political symbols—not just for communicating values, but also for distinguishing one's self from the opposition. These sorts of symbols are particularly effective as political tools because they are difficult to co-opt by the other side. It is harder for a Democrat than for a Republican to sound sincere when arguing for a strong national defense. As a matter of ideology, Democrats are often skeptical about the extent to which new weapons systems or new military adventures truly advance the cause of national security or human well-being. And similarly, it is harder for a Republican than a Democrat to sound sincere when arguing for the importance of science. Scientific results are commonly used to bolster arguments for government regulatory programs and policies, and as a matter of ideology Republicans are often skeptical about the ability of government to wisely design and implement such policies or about their actual benefits to society.

Neither of these ideological proclivities amounts to being, respectively, "soft on defense" or "anti-science," but each provides a nucleus of plausible validity to such accusations. Trying to go against this grain—as when Michael Dukakis, the 1988 Democratic presidential candidate, sought to burnish his defense credentials by riding around in a tank, or when George Bush repeatedly claimed that he would make decisions about climate change and the environment on the basis of "sound science"—inevitably carry with them the aura of insincerity, of protesting a bit too much.

And so perhaps we have now discovered the rightful place of science: not on a pedestal, not impossibly insulated from politics and disputes about morality, but nestled within the bosom of the Democratic Party. Is this a good place for science to be? For the short term, increased budgets and increased influence for the scientific-technological elite will surely be good for the scientific enterprise itself. Serious attention to global environmental threats, to national energy security, to the complex difficulties of fostering technological innovation whose economic outcomes are not largely captured by the wealthy, are salutary priorities of the

Obama administration and welcome correctives to the priorities of his predecessor.

Ownership of a powerful symbol can give rise to demagoguery and self-delusion. President Bush overplayed the national-security justification in pursuit of an ideological vision that backfired with terrible consequences in Iraq. In turn, a scientific-technological elite, unchecked by healthy skepticism and political pluralism, may well indulge in its own excesses. Cults of expertise helped bring us the Vietnam War, the recent financial crisis and many other setbacks and disappointments. Uncritical belief in and promotion of the redemptive power of scientific and technological advance is implicated in some of the most difficult challenges facing humans today. In science, the Democratic Party appears to have discovered a surprisingly potent political weapon. Let us hope they wield it with wisdom and humility.

ABOUT THE AUTHORS

Michael Crow
Michael M. Crow has been the President of Arizona State University since 2002. He is founder of the Consortium for Science, Policy and Outcomes. Prior to joining ASU, he was executive vice provost of Columbia University, where he also was Professor of Science and Technology Policy.

Robert Frodeman
Robert Frodeman is Professor of Philosophy and founding director of the Center for the Study of Interdisciplinarity at University of North Texas. His work ranges across environmental philosophy, the philosophy of science and technology policy, and the philosophy of interdisciplinarity. Frodeman's *Sustainable Knowledge* will be published in 2013.

David Guston
David H. Guston is Professor of Politics and Global Studies and Co-Director of the Consortium for Science, Policy and Outcomes at Arizona State University. He is also the Principal Investigator and Director of the NSF-funded Center for Nanotechnology in Society at ASU. His book, Between Politics and Science (Cambridge U. Press, 2000) was awarded the 2002 Don K. Price Prize by the American Political Science Association for best book in science and technology policy, and his papers are among the most well-cited in the field. Guston is also a fellow of the American Association for the Advancement of Science.

Carl Mitcham

Carl Mitcham is Professor of Liberal Arts and International Studies, Director of the Hennebach Program in the Humanities, and Co-Director of the Ethics Across Campus Program at the Colorado School of Mines. Publications include Thinking through Technology: The Path between Engineering and Philosophy (1994), Encyclopedia of Science, Technology, and Ethics (4 vols., 2005), Humanitarian Engineering (with David Muñoz, 2010), and Ethics and Science: An Introduction (with Adam Briggle, 2012).

Daniel Sarewitz

Daniel Sarewitz is co-director of the Consortium for Science, Policy and Outcomes, and Professor of Science and Society, at Arizona State University. His work focuses on revealing the connections between policy decisions, knowledge creation and innovation, and social outcomes. His most recent book is *The Techno-Human Condition* (MIT Press, 2011, co-authored with Braden Allenby). He is also a regular columnist for *Nature* and the author of *Frontiers of Illusion: Science, Technology and the Politics of Progress*.

G. Pascal Zachary

G. Pascal Zachary is the author of *Endless Frontier: Vannevar Bush, Engineer of the American Century* and *Showstopper*, about the making of a software program. He is a Professor of Practice in the Consortium for Science Policy and Outcomes and his articles on the politics of technology and science have appeared in *The New York Times, Spectrum, Wired* and other publications. Zachary is editor of The Rightful Place of Science series.

ACKNOWLEDGEMENTS

The essays in this volume represent some of the wisdom accumulated by their authors prior to and during the first decade of the 21-st century. All but two of the essays were conceived with the encouragement and guidance of Kevin Finneran, editor of the National Academy of Science's quarterly journal, *Issues in Science & Technology*. Finneran and *Issues* generously allowed these essays to be published here in somewhat different form. Here are the original titles and dates of publication of those essays:

"Retiring the Social Contract for Science," Summer 2000, by David Guston

"Beyond the Social Contract Myth," Summer 2000, by Robert Frodeman and Carl Mitcham

"Forget Politicizing Science. Let's Democratize Science!," Fall 2004, by David Guston

"Does Science Policy Matter?," Summer 2007, by Daniel Sarewitz

"None Dare Call it Hubris: The Limits of Knowledge," Winter 2007, by Michael Crow

"The Rightful Place of Science," Summer 2009, by Daniel Sarewitz

A different version of the essay, "Technology and Social Transformation," appeared in an annual survey from the American Association for the Advancement of Science. An even earlier version of the essay appeared in a report from a workshop on Nanotechnology and Social Transformation, sponsored by the National Science Foundation and held in September 2000.

"Power and Persistence in the Politics of Science" was written by Michael Crow and Daniel Sarewitz specifically for this volume. Special thanks to them for their original contribution.

Having gone fully digital in recent years, the publishing world contains new vistas for both independent thinkers and innovative institutional actors. Travis Doom, a staff member in the DC office of the Consortium for Science Policy & Outcomes, has proved essential. He managed the many aspects of producing this volume. Suzanne Landtiser designed the superb cover and provided additional design guidance. Lori Hidinger, the chief operating officer of CSPO, lent her expertise to the project at every stage. Thanks to all.

In conceiving of the Rightful Place of Science series, and this volume in particular, I received help from each of the contributors to this volume but especially from David Guston, co-director of CSPO. Guston's combination of exacting scholarship and passion for literary quality aided this project throughout. I also wish to thank Clark Miller, professor of politics at ASU and CSPO's academic lodestar.

I also received guidance and encouragement from John Markoff, a science writer at *The New York Times*. A discussion with Nancy Toff, a senior editor with Oxford University Press and the American editor for its *Very Short Introduction* series, proved especially beneficial. The Oxford series provides inspiration to me and evidence that short paperback originals, clearly written by leading scholars, but lacking scholarly apparatus, can and do find an enthusiastic audience and a global one. Finally, I'm grateful to Liam Zachary, my son, who tutored me on new forms of book publishing.

-G. Pascal Zachary